Water, Power and Persuasion

How Jack Pfister shaped modern Arizona

Water, Power and Persuasion

How Jack Pfister shaped modern Arizona

by Kathleen Ingley

AMERICAN
TRAVELER PRESS

Printed in the United States of America
2015 printing

ISBN: 978-1-55838-178-0

American Traveler Press
5738 North Central Avenue
Phoenix, AZ 85012
800-521-9221
www.AmericanTravelerPress.com

Editing: Bob Albano, Phoenix, Arizona
Cover/Interior Design: The Printed Page, Phoenix, Arizona

Contents

Foreword

By Bruce Babbitt
Former Governor of Arizona and U.S. Secretary of the Interior

Arizona water management was a chaotic jumble of mismatched and unworkable parts when Jack Pfister was named general manager of Salt River Project in 1976. The structure was poised to collapse.

Jack played a key role in rebuilding and modernizing how the state handles water. In the process, he emerged as Arizona's foremost public citizen.

The state's water challenge became a crisis in 1977, when President Jimmy Carter withdrew federal support for Orme Dam, the principal flood-control feature of the Central Arizona Project, then under construction. And by that time, the Arizona Supreme Court, never a voice of clarity on water issues, had arrived on the scene with a series of decisions upending our antiquated groundwater laws.

Then, in 1978, a three-year cycle of rainstorms began to overwhelm the storage dams on the Salt River above Phoenix. Flood crests poured through Central Phoenix again and again, cutting the city in two.

Salt River Project stood squarely in the intersection of these accelerating events. From its beginning as the first big, successful federal reclamation project, SRP had come to dominate Arizona water management. Yet across the years, as it promoted the interests of its agricultural constituency, it gradually lost touch with the needs of Arizona's emerging urban economy.

Jack took leadership of Salt River Project at the worst of times, or perhaps at exactly the right time to begin a career of reshaping both that institution and Arizona's water-resource future. And it began with Orme Dam.

President Carter's decision to cancel Orme Dam triggered a near-hysterical reaction from the Arizona congressional delegation and editorial writers at the *Arizona Republic*. Opponents of the dam, however, including the Fort McDowell Yavapai Nation, stood firm in opposition, steadily gaining allies both in Arizona and in Washington.

As the controversy escalated, without an obvious resolution in sight, I, as governor, appointed a seventeen-member committee that included both supporters and opponents, and charged it to search for a consensus solution. After more than two years of public hearings, extensive information outreach, and technical studies led by the Bureau of Reclamation, the committee reached a nearly unanimous conclusion: Orme Dam was not necessary. An alternative that came to be known as Plan 6 would be adequate for both water storage and flood control.

In the fall of 1981, as the committee began to coalesce in favor of Plan 6, Jack then emerged from the background to release a letter from Salt River Project addressed to the committee. He began by reiterating SRP's support for Orme Dam, clearly intended to keep the peace with the pro-dam farmers who dominated his board of directors. However, he then proceeded to analyze the other alternatives formulated by the Bureau of Reclamation in a manner that effectively undercut, indeed demolished, the case he had just made for Orme Dam as the non-negotiable establishment preference.

Shortly thereafter, the committee voted to drop Orme Dam in favor of Plan 6. The old guard struck back, and the political fight commenced. Jack was out on a limb but not about to back away.

He then persuaded me to accompany him to Washington to brief members of the Arizona delegation. We were greeted as interlopers with no business questioning the delegation's wisdom.

As we left town, I suggested we should take on the delegation in public. Jack shrugged, unperturbed, and responded, "Governor, that would not be a good idea. No need to get emotional. We have the facts." As usual, it was good advice. In due course, the congressional delegation came around, Arizona came together, and today Plan 6 has been largely implemented. The Salt River dams have been upgraded to modern flood-control standards, and the Central Arizona Project is complete.

Jack Pfister with Gov. Bruce Babbitt

In the meantime, another piece of our water-management structure had collapsed. The overdrafting of our groundwater basins had been a controversial issue for half a century. By the late 1970s, it had reached a crisis point as the result of court decisions described in this book. Successive legislative committees and study groups ended in deadlock. Finally, I was invited to give it a try.

I gathered together yet another committee. Consulting with Jack, I decided on a different approach from the highly public Plan 6 process. We would invite just a small group of establishment insiders, representing the three warring factions: cities, the mining industry, and agriculture. We would meet in private outside the Capitol, and participants would agree not to talk in public or to the press. It was time for deal-making, and transparency and public process went out the window.

The outcome of those negotiations, the dramatic enactment of the Groundwater Management Act of 1980 is an oft-told story. Less appreciated is how Jack Pfister managed to make it possible.

Meaningful groundwater management would be a direct challenge to the primacy of Salt River Project, something his board would be unwilling to support. Jack would have to play a hidden hand, first by staying away from the table, instead sending a trusted member of his staff. And second by encouraging me to keep other farm representatives out of the process, so as to avoid stirring up further controversy. To save agriculture, we had to shut their leaders out.

With the process arranged this way, it was then possible for Jack to begin making concessions, capably described in this book, that would be necessary to reach agreement, including the creation of a state Department of Water Resources to administer the resulting regulatory system. The Groundwater Management Act of 1980 did little for Jack's standing at Salt River Project, but in the long run, it has worked to the immense benefit of all Arizonans, including the agricultural constituency of SRP.

It was from these two episodes, the Orme Dam controversy and passage of the groundwater code, that Jack became Arizona's chief consensus builder and civic activist.

There was more to come, including contentious negotiations over the state's share of the bill for the Central Arizona Project and dam safety, the passage of a strong water-quality law that also created a state Department of Environmental Quality, and the establishment of a legal framework for groundwater-recharge programs.

As these projects moved forward, my partnership with Jack came full circle as he assumed public leadership and moved to the head of the table to lead negotiations through the various hybrid committees we created.

During our time together, we worked together on many other matters, described in engaging detail in the following pages. This book is much more than a biography. It is a penetrating look at the art of transformative leadership, of how one person can change the course of public events, in the process becoming the most productive and influential public citizen of his generation.

—Bruce Babbitt, May 2015

Why This Book Was Written

Many people knew that my father was working on a political biography of former Majority Leader Burton Barr, and although he had done some wonderful research and logged some great oral histories (check out www.BurtonBarr.com to hear them), he was not able to finish the book due to his untimely death. So I felt it was part of my family's duty to see that the Barr book was indeed finished.

As we worked through the process of getting that book completed, other people came to me and said, "So when is the book about your father going to be written?" I heard countless stories of how he had mentored people, and helped them in both small and large ways. They recited his "Pfisterisms," they talked about how he had made time for them, and how they knew they could always count on him to get things done. He was the ultimate "servant leader," and showed it in so many ways that we simply felt we had to share his story.

I thank Kathleen Ingley for all her hard work in making his life and accomplishments come alive. No great leader stands alone, and my dad would be the first to say how much other people supported him while he was serving the community. But it was his unending curiosity for learning and genuine interest in people — and his love for the state of Arizona — that propelled him to keep on trying to make it better.

Oh, that we all could take a page out of that playbook!

—Suzanne Pfister, September 2015

Acknowledgments

The person who helped me most in writing this biography is, sadly, not here to thank. Jack Pfister's well-organized stash of clippings, letters, and memorabilia was a treasure. (The one lapse was the failure to identify most of the photos — a lesson for the rest of us.) I knew Jack as a source for news stories, a sounding board for political realities, and the fellow member of a book group. But researching his life brought a steady stream of surprises as I kept turning up groups he'd helped, issues he'd influenced, and people he'd mentored. The list no doubt goes beyond those I discovered.

The Pfister family generously shared their memories and Jack's extensive files. Thank you to Suzanne, Scott, Tad, and Pat.

I'm deeply grateful to those who shared their perspective on Jack's life. The reality of time forced me to limit the number of interviews — I could easily have spoken to three times as many people with valuable insights. The reality of space forced me to leave out, with great regret, many of the quotations and stories from those I did interview. Whether or not they're cited in this book, all contributed to this portrait of Jack's life: (To read additional memories of Jack and to share your own, go to www.JackPfister.com)

Paul Ahler, Carolyn Allen, Edith Auslander, Brittany, Bryant and Loretta Avent, Bruce Babbitt, Sandy Bahr, Maria Baier, Betsey Bayless, Mike Bielecki, Mark Bonsall, Molly Broad, Jack DeBolske, Carolina Butler, Ernest Calderón, José Cárdenas, Dan Campbell, Joe Cayer, Herman Chanen, Michael Clinton, John Christian, Arlan Colton, Cathy Connolly, Lattie Coor, Bill Davis, Dino DeConcini, Saul Diskin, Cathy Eden, Nan Ellin, Paul Eckstein, Paul Elsner, Dr. Paul Eppinger, John Fees, Kathy Ferris, Grady Gammage Jr., Phil

Gordon, Pat Graham, Athia Hardt, Chris Herstam, Elliott Hibbs, Bill Hogan, C.A. Howlett, Jane Hull, Andrew Hurwitz, Paul Johnson, Edib "Ed" Kirdar, Ruthann Kizer, Rabbi Robert Kravitz, Jon Kyl, Lisa Loo, Stanley Lubin, Bob Mason, Leroy Michael, Dennis Mitchem, Ioanna Morfessis, Nancy Neff, Luther Propst, Al Qöyawayma, Mike Rappoport, Bob Rink, Steve Roman, Dr. Marty Rozelle, Elisabeth Ruffner, Claire Sargent, Karen Scates, Bill Schrader, Dan Shilling, Herbert Schneider, Bill Shover, Martin Shultz, Richard Silverman, Karen Smith, Rob Smith, Dr. Warren H. Stewart Sr., J. Fife Symington III, Mollie Trivers, Kim VanPelt, Carl Weiler, Richard Wilks, Sid Wilson, Mack Wiltcher, Dr. Robert Witzeman.

Salt River Project provided extensive help in tracing Jack's career there. Particular thanks to General Manager and CEO Mark Bonsall, Associate General Manager and Chief Communications Executive Gena Trimble, Senior Historical Analyst Catherine May, and Media Relations Manager Scott Harelson.

Editor Bob Albano spotted glitches, fixed style lapses, and led the tough job of cutting. Publisher Bill Fessler and designer Lisa Liddy handled the magic of turning a manuscript into a book.

Chuck Kelly and Richard de Uriarte, my former colleagues at the *Arizona Republic,* generously gave advice and guidance. A huge thanks to them and all the other journalists who write about the unglamorous subjects, such as water and education policy, that have an enormous impact on our daily lives. They dig up the facts and insight we need to make informed choices about the future. But they're an endangered species as changing technology erodes the media commitment to this foundational coverage.

Writers probably shouldn't admit when words fail them, but it's impossible to adequately express my gratitude to my husband, Howard Seftel. He gave me boundless support, endless patience in looking at drafts, and, as a fellow writer, invaluable advice on wording and structure. (Yes, it's OK to split infinitives.)

—Kathleen Ingley, September 2015

Introduction: The Change Agent

Stewart Mountain Dam was about to burst. It was February 1980 and two huge storms had clobbered Arizona. A third was on the way. If that storm hit full force, water would rise to the top of the aging dam forty-one miles upstream from Phoenix, and it would crumble. Water would roar through the city and its suburbs: up to three million gallons a second. It would pour into businesses, homes, and the Capitol complex. And two hundred thousand people would have to be evacuated.

Arizonans "must be prepared for the unthinkable," the governor warned in a late-night press conference on February 15.

No one knew the risks better than Jack Pfister. As the top executive of Salt River Project, he headed the utility that operated the string of dams around Phoenix. He knew the structures were designed for water storage, not flood control.

SRP was already under attack for flooding a swath of the Phoenix metro area. A relentless series of storms had forced dam managers to release massive amounts of water down the normally dry Salt and Verde rivers. The water had taken out bridges, washed out river crossings, inundated neighborhoods, and flooded part of Sky Harbor International Airport. Phoenix had racked up millions of dollars in damage and lost business.

By luck or a miracle, that last storm veered away. Stewart Mountain Dam held. But Jack Pfister knew that Arizona would have to change how it handled water. Neither the state nor Salt River Project could go back to business as usual.

<>〈〉<>

Born in 1933, Jack grew up in an Arizona with a small-town vision of itself, with the unbridled confidence that the only thing better than growth was more growth. It was a place of boosters and hucksters and new arrivals bent on remaking themselves. As every kid learned in school, the state's economy was based on the four Cs: cattle, cotton, citrus, and copper — with the climate that attracts tourists sometimes added as number five. Rural lawmakers exerted outsized leverage at the Legislature.

But the state's self-image was becoming less and less like reality by the 1940s. Arizona was urbanizing rapidly, and the pace picked up dramatically after World War II. Jack Pfister became a key figure in bringing Arizona and Salt River Project into the modern era. Until his death in 2009, he had a hand in confronting many of the state's toughest challenges, from ensuring a clean and stable water supply to managing explosive population growth.

As the general manager of Salt River Project from 1976 to 1991, he led a powerful institution begun in 1903 as a not-for-profit association to get federal loans for dams. Shortly afterwards, the association took advantage of the potential for hydropower and became a major supplier of electricity. Salt River Project is now the nation's second-largest public power utility, based on generation, and has the fourth-largest customer base.

Jack helped transform it from an organization narrowly focused on agriculture to one that took part in major issues like the environment and social justice. He modernized its management, moving from authoritarian to participatory style.

Besides his role with Salt River Project, Jack left his mark on higher education as a member of the Arizona Board of Regents, Arizona State University staff, and the Maricopa Community Colleges Foundation. He recruited a university president, worked to get more minorities into college, wrestled with tuition increases, and defied a governor over free speech.

He made a point of reaching across religious, racial, and ethnic lines. "States that are internationally competitive," he observed, "will be those that provide cordial environments for diverse populations and draw their strengths from their multicultural society."[1]

"For reasons that I don't understand, I am not afraid to expose myself to new ideas," Jack told *Phoenix New Times* in 1990. "As a matter of fact, I find it invigorating. And I have made many significant changes in my views based upon a growing body of evidence that I was in error."

"Jack was very much a forward thinker," said former Arizona Sen. Jon Kyl. "He was at the leading edge of change in a lot of things in the community and at Salt River Project. Jack was a change agent. He challenged old ways of doing things if he thought they needed to be modernized."[2]

Jack was a quintessentially American success story, losing his father at the age of three, raised by a single mother, and rising to become one of the most influential figures in the state. He had a Westerner's confidence in progress and a Ben Franklin's belief in the power of self-education and self-improvement.

Jack developed such clout and respect that his name was tossed around as a possible candidate for governor during the 1987-88 recall push against Gov. Evan Mecham (who was sent packing by impeachment, instead). Jack brushed off the suggestion. Why should he have political ambitions, insiders quipped, when he already had as much power and influence as the governor?

You'd never know Jack's status to look at him. He wore boxy, nondescript suits that did nothing to flatter his well-rounded proportions. His thick glasses had sturdy, practical frames that made him appear both owl-like and approachable. He preferred khakis to jeans and wore an eclectic succession of hats outside to protect a balding head from the ferocious Arizona sun.

Jack wasn't interested in wealth. He lived for four decades in a low-slung, stucco ranch house in a comfortably middle-class part of Phoenix. For years, a neighbor, who had never met the family, assumed that the man who left for work every morning with a bulging briefcase was some sort of junior accountant. He was stunned to learn he'd been watching one of the top executives in Arizona.

Jack was the go-to person if you had a tough issue of public policy. Serving on countless boards, committees, and commissions, he was a driving force in everything from creating a long-overdue state Department of Environmental Quality to getting voter approval for a Martin Luther King Jr. Day holiday after a storm of controversy.

His secret: listening, truly hearing what each person had to say. At endless meetings, Jack paid attention to each speaker while others in the room were looking out the window and counting the planes going by. Jack knew how to digest what he'd heard. He had a genius for spotting common ground and figuring out compromises.

Although Jack was a lifelong Republican, he never thought the party had a monopoly on good policies. He served on the transition teams for Gov. Jane Hull, a Republican, and Phoenix Mayor Phil Gordon, a Democrat.

While his fingerprints are on a wide variety of changes, they're sometimes hard to find. He stayed behind the scenes to help craft Arizona's historic Groundwater Management Act. Friends and colleagues knew that Jack was involved somehow in dealing with the aftermath of Three Mile Island, when a partial meltdown exploded the public's confidence in nuclear power. But his name isn't on the big committee reports. Instead, he was the crucial, unheralded intermediary in creating an organization that would standardize and ensure safety measures.

Jack focused on accomplishment, not accolades. If that meant letting others get most of the credit, it was fine with him.

Most of all, Jack loved being a mentor. He was always ready to give advice over lunch or listen sympathetically over a cup of coffee. Arizona used to be run by a few elite movers and shakers, including him. But he saw that power was becoming more diffuse, and he welcomed the change. He called for "a large number of people who are willing to take risks and assume leadership roles."

Jack set an example of public service that few can match. His results-oriented approach to tough issues — respecting differences and looking for common ground — is more relevant than ever. Jack had a passion for his home state and its colorful past. He had a couple of projects for books that he never managed to finish. It's no wonder he couldn't find time to write about Arizona history. He was too busy making it.

Prescott Roots

The little boys crawled under the wire floor of the turkey coop. Keeping their heads down, they tried to ignore the stench while they scooped up piles of droppings.

Jack Pfister and his younger brother, Tad, were still in elementary school when they had their first job: cleaning a neighbor's turkey pens.

Life should have been a lot easier for them. They were born to a hard-working father with a drive to succeed: Alfred John Pfister managed to build a thriving service station business in the middle of the Great Depression.

The son of Swiss immigrants, Al was working by age eleven. He was born in 1904, either in New Orleans or, some records say, Mobile, Ala., where his three older siblings were born. When his brother, Herman, developed tuberculosis, the family followed the standard treatment of the time: move to Arizona. Prescott, semi-arid and a mile high, was a logical choice. An old Army post there, Fort Whipple, had been reactivated as a tuberculosis sanatorium and then a hospital for disabled veterans.

Al and his mother moved to Prescott along with Herman in 1924. The two sisters, Sophie and Lillian, and their families came along, too.

Back then, they spelled the family name "Pfiester." Sometime over the next six years, they dropped the "e," but kept the long-e pronunciation (*fee-ster*).

Here was a picture-book American small town of red-brick commercial buildings with awnings and charming wood homes with inviting porches. (It still is, in many ways, with the official website proclaiming, "Welcome to Everybody's Hometown.") Prescott is the county seat, and Yavapai County Courthouse, with imposing pillars on all sides and a facade of local granite, dominates the shady square in the center of town.

Al and Herman worked for a time at the Whipple health complex. But they soon struck out on their own, buying a one-pump gas station six blocks east of the courthouse, on the corner of Gurley and Washington. The family lived in a duplex behind it.

But Herman wasn't healing in the dry climate, and Al had to take the lead in running the service station. Around 1930, he became an agent for the Texas Company, which later shortened its official name to Texaco.

Two years later, Al married a twenty-seven-year-old local girl, Roberta McDonnell. Everyone called her Bobby, and she had a flair for performance — a talent that later proved crucial in keeping the family afloat.

Bobby came from tough stock, a family with women who knew how to survive on their own. Her grandmother, Lizzie Klar, was born in 1859 into a prosperous Chicago family with a furniture factory. But the Great Chicago Fire destroyed the business in 1871, and Lizzie's father became a carpenter paid by the day.

Lizzie married a local lawyer and gave birth to a boy and a girl. But her comfortable circumstances didn't last. Her husband took off for San Francisco and she had to go to work.

Then, when the children were in their teens, her life took a fairy-tale turn. An old sweetheart, Robert E. Morrison, was offered the post of U.S. attorney in the Arizona Territory, and on his way to Washington to accept, he stopped in Chicago to see old friends. Morrison was a widower, and when he saw Lizzie, the spark was still there. She submitted documents to the Catholic church showing her first husband was dead, and the pair married in 1898.

Lizzie's son was grown and established in Chicago — he started working at thirteen to supplement his mother's meager income. So only her daughter, fifteen-year-old Inez, came along on the move to Arizona. The Morrisons built a house in Prescott, and the family story is that Lizzie helped design it.

Two strong women faced off when Inez fell in love with a doctor after graduating from high school. Lizzie didn't approve of John McDonnell. Inez sneaked out to meet him anyway, and the two eloped. They moved to the mining town of Jerome and had three daughters. Bobby was the oldest, born in 1905.

The popular young doctor had an attack of appendicitis in 1911. He had surgery, but died a few days later. He was just thirty-six years old. Inez moved back to Prescott, where her mother could help with the three little girls.

Now Bobby was grown and showing off her talents and organizational skills. She and her sister Betty organized the "Whipple Revue," a series of chorus numbers and skits performed by male hospital patients and personnel in drag at the medical site. It was such a hit that it was booked to play before the regular Saturday night movies at the Elks theater, with a bump in ticket prices to raise money for the American Legion auxiliary. The local newspaper pronounced it "a scream and a success."

Around 1925, she directed the Holly Follies, a fundraiser for the municipal Christmas tree fund that had the chorus throwing cotton snowballs into the audience and, a reporter noted, dancing together "in perfect time and unison."[1]

But Bobby's ambitions were frustrated. She went to the University of Arizona in Tucson, then couldn't afford to stay. She taught kindergarten for several years, between 1926 and 1929 — more to bring in money, it seems, than out of any special interest in the job.

The outgoing, accomplished young woman caught the eye of young Barry Goldwater, the future senator and presidential candidate. Although he grew up in Phoenix, his family owned a store in Prescott and had a home near Bobby's.

The two began dating, and Goldwater later described her as the love of his life at the time. While the romance fizzled, they remained friends, and years later, when the Goldwaters came to a cabin near Prescott, Jack would sometimes play with Barry's children.[2]

But there was one part of Bobby's early life that her sons wouldn't discover for decades: She had been married before she wed Al in 1932. After she died, Jack checked divorce records in Yavapai County and learned the name of her first husband. Frederick Wall was a salesman traveling through Prescott when Bobby met him. He'd lived in Bisbee in southern Arizona and then moved to San Francisco. Bobby, perhaps to follow him, enrolled in a summer library course at the University of California at Berkeley in 1930.

Wearing a blue chiffon outfit and a matching hat, she married Wall on July 19 in the Catholic chapel near campus.

It sounded so romantic in the newspaper wedding announcement: They were going to honeymoon in Carmel-by-the-Sea, take an apartment in Berkeley while she finished the summer term, and then move to San Francisco.

Reality was harsher: Barely six months after their vows, Wall walked out. Bobby got a divorce in Yavapai County, granted on April 19, 1932. The grounds were desertion and abandonment.

Al filled the gap in her life. Because Arizona prohibited remarriage within one year of divorce, Bobby and Al drove over the state line and were married in Gallup, N.M. — two weeks after the divorce decree.

They were a striking couple. Al was Hollywood handsome in 1930s style. With smooth dark hair and soft features, he looked like the faithful best friend in a musical comedy. Bobby — lithe, with broader features and a lock of marcelled hair brushing her forehead — could have played a wisecracking girl on the chorus line.

Their marriage, though, was touched with the sorrow of recent deaths. Herman had died in September 1931. Then, on Christmas Eve, Al's mother went into the hospital. A week later, on January 2, 1932, she was dead at sixty-three.

Alfred John Pfister Sr. holding his infant son, Jack, 1933

But the next year brought new life. Bobby was pregnant. A son was born on October 3, 1933. The family lore is that during the pregnancy, Al had a few too many drinks and promised his mother-in-law, Inez, that she could name the baby — a pledge that he immediately forgot.

A few days after the birth, she went to the hospital, held the baby, and announced, "I'm really pleased with little Jackie."

No, the family said, his name is Alfred John.

"No," she shot back, "you promised me that I could name him, and I'm going to name him after my own husband, whose name was Jack McDonnell."

No, said the family, the birth certificate was already submitted with the name "Alfred John."

"Well, I'll call him Jack," she said. And it stuck.[3]

Jack got a brother two years later, on May 5, 1935. The baby was named Herman Frederick, after his late uncle. As another family story goes, Jack couldn't manage the name and called his brother "tad," maybe meaning to say "tot." That nickname stuck so thoroughly that virtually no one ever knew his given name.

Both Al and Bobby had lost their fathers early in life. Al was just two when his father died at forty-two. Al used to joke that he would go even sooner, dying before he reached forty, his mother-in-law recalled. Everyone kidded him about the dire prediction.

But Al took the premonition seriously enough to take out ample insurance policies. They included coverage to pay for the boys' future college education: $10,000 for Jack and $5,000 for Tad. The foresight turned out to be vital.

Meanwhile, though, business was looking up. In mid-1933, Al retired from the retail side so he could concentrate on wholesale operations for the Texas Company. He leased out his Prescott station. He bought property in nearby Granite Dells with cabins, a store, and a gas pump. His two sisters and their husbands helped out.

Even in the middle of the Great Depression, real financial success was just around the corner for Al. In late 1936, he collected the annual rent for his garage and used part of the money for insurance premiums. He set the rest aside to take his family and the maid to the Sugar Bowl game in New Orleans. Maybe, he said, he'd get a new car for the trip, too.

And then he came down with pneumonia. He went into Mercy Hospital on December 21. By Christmas Eve he was in an oxygen tent. "He had all the medical attention a millionaire could have gotten," his mother-in-law wrote later. But on the twenty-ninth, he went into convulsions and died. His premonition was right — the four-decade mark was eight years away.

Now Bobby had to worry about money. The insurance provided only a short-term cushion. Al had extended a lot of credit to Depression-squeezed customers. He was also in debt to the Texas Company, which ended up with all the assets except the Prescott service station and the house next door.

And for all his foresight, Al died without a will. His two sisters — who had "bled him for years & nearly driven him crazy," according to Bobby's mother — saw a chance to cash in.[4] Their husbands sued the estate for back wages and loans. Fortunately, Bobby had the legal ammunition to defend herself. Her sister Betty had married a talented local attorney, Ed Locklear, who fought off the claims. But the battle created a bitter rift in the family that lasted a half-century. Bobby and Al's sisters cut off relations so thoroughly that there was no contact between the two families until 1988, when one of Jack's cousins got in touch with him.

The lease on the service station didn't bring in enough money to live on. Then Bobby found an opportunity that fit her talents. In 1940, local businessmen built a radio station to serve Prescott with local news and music. Four years later, KYCA — the call letters stood for "Yavapai County Arizona" — became the 187th affiliate of the National Broadcasting Company.

Bobby became program director at KYCA and had her own show, "Bobby Pfister's Briefcase," with household hints and Hollywood gossip. "She'd read recipes, and her family laughed gales, because she never cooked," recalled Elisabeth Ruffner, who lived up the street from Bobby and the boys.[5]

The job, with her power to grant air time and mention local events, connected Bobby with a wide variety of organizations. The thank-you notes to her ranged from the Salvation Army to the Army Air Forces to Yavapai Associates, a coalition of county civic groups.

Bobby had company as a single mother. Elisabeth Ruffner was on her own, too, since her husband had enlisted in the military. With different schedules — Bobby didn't go into work until noon, while Elisabeth worked at a doctor's office — the two women watched out for each other's children.

Bobby dazzled her twenty-year-old neighbor. "To me, she was the most glamorous, fabulous, intellectual woman," Ruffner said. "She smoked and drank and wore makeup."

Bobby was also exceptionally kind. When the Ruffner toddler developed asthma, Bobby would come over and rock her at night.

Bobby broke through the prejudice barriers against Mexican-Americans, Ruffner said. To introduce her show, she hired guitarist Augustine Rodarte, who had come to Prescott from Zacatecas, Mexico.

At a time when casual bigotry was the norm, Tad Pfister said, his mother insisted that they should never discriminate against anyone.[6] Whether it was in those early years or later, Jack developed a passion for tolerance and a commitment to minority rights and access to education.

A World of Learning

Prescott was a great place to be a kid. Jack and Tad lived across the street from the town stadium and park, now called Ken Lindley Field and Park, and they treated the place as a giant backyard. The family would go there for community sings, with the words to songs projected onto a wall. Jack became batboy for a number of softball teams in an era when the sport was a big deal in town.

The brothers would cut willow branches, light the end on fire, and stick them into wasp nests in crevices in the stadium wall. They'd always end up getting stung and go home crying, where Bobby had baking soda for their stings but very little sympathy for their pain.

Bobby, Tad, and Jack sitting on the front porch of their Prescott home, 1944

Every Saturday morning, Bobby would give each a quarter to go the movies — fifteen cents for a ticket and ten cents for candy.

The two boys would go swimming in the reservoir at Granite Dells, the fancifully eroded rock formation outside Prescott. Their mother took them picnicking, fishing, and hiking. They learned to shoot guns. When they joined Cub Scouts, Bobby was a den mother. Jack went on to Boy Scouts, earning merit badges that included path finding, home repairs, swimming, and woodworking.

Believing in education and hard work, Bobby made sure the boys did too. When they were little, she read to them every night. For every wish they came up with — from getting a bike to playing golf — she had one answer: Go to school and get an education. She was determined that her sons would get the college degree that had been out of her reach. Years later, after starting veterinary school in Colorado, Tad wrote to tell his mother he was struggling. She shot back a postcard with a stern message: If you flunk out, don't come west; go east.

Jack and Tad were maybe ten and eight when they got the job cleaning turkey pens. That was the start of a string of jobs, from delivering newspapers to cutting Christmas trees to shoveling snow. If they couldn't find work, their mother or Uncle Ed would line up something.

Bobby didn't talk about money, but finances were clearly tight. When the Sears Christmas catalog came, Jack and Tad joined thousands of other American kids in going through the toy section page by page. They eagerly took turns putting dibs on their favorites — even though in the end, they knew they'd get just one gift for Christmas.

Fortunately, Bobby had a banker who would lend her money to get over rough patches. He also helped her with financial management, not her strong point. It was Jack who saved up and bought his brother a bicycle, Tad remembered.

The family had deep roots in Prescott on Bobby's side and through their uncle, so there were a lot of people to look out for the two fatherless boys. "We were kind of raised by the town, so to speak," Jack explained in an interview.[1] Uncle Ed, married to Bobby's sister, made sure his nephews stayed out of trouble. As a prominent attorney, he was a role model of achievement, and he pushed them to excellence with demanding standards. But he wasn't an affectionate man, and they got no paternal warmth from him.

Bobby had been brought up Catholic and made her sons go to Mass on Sunday. But she never set foot in the church herself: She'd been excommunicated. The Catholic Church didn't recognize civil divorces. So Bobby was considered an adulteress and a bigamist. For Americans, remarrying after valid Catholic marriage was grounds for excommunication. That draconian penalty, set by the Third Plenary Council of Baltimore in 1884, wasn't revoked until 1977.

The local church wasn't charitable about Bobby's status, according to a story Jack told his daughter. As a little boy, he came home from church one day and asked, "Mommy, what's a 'bastard'?"

Where did he hear that word? Bobby demanded.

From the priest, Jack answered. "He said Tad and I are bastards."

The two brothers had totally different personalities. Tad was a natural schmoozer, a smooth dancer, a ladies' man, and the one who got into mischief. Jack was a classic first child, obedient and studious. While Tad was naturally lean, Jack never lost his early pudginess and would struggle with weight his whole life.

At Washington Elementary School, Jack found a role model in the principal. Sinclair Louttit was one of the many men in town who kept a kindly and watchful eye on the fatherless boy and his brother. Louttit, who had lost his right arm in a hunting accident, was certainly an example of the can-do spirit and optimism that Jack would show throughout his life.

Jack practiced penmanship with an old-fashioned ink pen, trying to make his ovals as perfect as possible. When he was complimented on his cursive writing later on, he realized that those apparently "painful and boring exercises were really of value."[2]

World War II seemed far away. Then, after D-Day, the teacher began putting pins on a map to chart the Allies' progress in Europe.

It was an era when schools staged annual Christmas pageants. As Jack's class got ready for its part in singing the carols, he recounted six decades later, "my teacher discovered my inability to carry a tune and forbade me from singing out loud at the event. To this day, I do not sing out loud."

Jack needed braces, but there were no orthodontists in Prescott. With the war, gas rationing limited how far anyone could travel. So some mothers pooled their gas stamps, and every month one of them would drive three or four kids to Phoenix to have their teeth checked.

Overweight, with glasses and those braces, Jack had been shy in grade school. Then, in junior high, he began playing football. It was a life-changer. While he wasn't particularly athletic, "he was strong and competitive," said friend and teammate Mack Wiltcher.[3]

Football gave him self-confidence, and he was on the varsity squad for three years. "I was never very good but I tried hard," Jack recalled. Then again, the team wasn't very good either. A tie game was the only reason it didn't have all losses his senior year.[4]

Jack was in school plays, too, although not, as Mack remembered, in speaking parts. And girls liked to date him, because he was fun.

Teenagers Tad and Jack, Prescott, 1948

"No one disliked Jack Pfister," Mack said. "He was just one of the most personable persons. He was cocky a little bit, but he wasn't pushy. You couldn't help but like him."

The Wiltchers had made their way to Arizona for the same reason as the Pfisters: health. Mack's family came in 1947 as his mother's last chance of surviving asthma. His father found a job in a speck of a place called Bagdad, and Mack had to stay in Prescott to finish high school.

The first year, when he was a junior, he "bached" it, sharing a room with a senior. By the time his roommate graduated, he'd become friends with Tad and Jack. His parents hit it off with Bobby, partly because of her late husband's Southern roots, Mack recalled. When senior year rolled around, Bobby suggested he come live with them.

The Pfister home had a separate bunkhouse out back, where a Mexican-American family was staying when Mack moved in. The wife, Lorraine, kept house for Bobby. Fortunately for Tad, Jack and Mack, the family moved out and Bobby fixed up the bunkhouse for them.

To feed three teenage boys with healthy appetites, Lorraine used to cook up a big bowl of Spanish rice. One day a friend came over for dinner, and someone lobbed a spoonful of rice. Soon they were in a Spanish rice fight that spread through the house. Lorraine cleaned up the mess — and didn't speak to the boys for a week. When Bobby came home, "I was afraid she was going to ship me out," Mack said. But she was willing to put up with mischievous boys.

One night, Jack was sitting reading a newspaper in the bunk-house. That looked far too serious to Tad and Mack. So they got a firecracker, about eight inches long, lit it, and rolled it into the room, Mack remembered. The explosion sent Jack tearing out of the room. While Tad and Mack were hooting with laughter, the trick turned on them. A spark landed on their bed and soon clouds of smoke were boiling out of the smoldering mattress. Bobby chewed them out but didn't make them pay for it.

"She was such a sweetheart," Mack said. And she knew how to listen. "When we had problems at school, we could go talk to Bobby. She would hear me out, she was a good counselor." It was a trait Jack would share, as well as Bobby's supremely practical search for solutions.

She was very logical, Mack found. She'd walk him and others through problems, especially their love lives. "We'd be moaning and groaning," he said. "She'd say, 'You've just got to settle down.'"

Mack and his wife-to-be, Scotty, would double-date with Jack.

He had a sense of humor with an impish side. Scotty's mother had an upright piano with a stool that spun. Once, Jack hopped up on the stool and began doing a version of the Charleston. Suddenly one of the chair legs snapped off. Jack's reaction? He "laughed himself to death," Mack said.

Already Jack's leadership skills were showing. Listening was his strength. "He would hear you out," Mack said. In high school and college, he would listen to someone to the end before responding. "I never did see Jack put anybody down," Mack said. And he was careful about credit. "He would never take credit for something you did. He would always make sure the credit was yours."

Bobby set an example for civic service. She narrated a local spring fashion show, supported American Education Week, and helped drum up business for local merchants on "$ Day." The war effort became a personal cause. She chaired the civilian Air Corps committee. While her brother-in-law, Ed, led the Victory Bond drive, Bobby headed up the women's side of the campaign. When Prescott was falling short of its goal in November, Bobby put twelve-year-old Jack on the air, shaming listeners with visions of injured soldiers — missing an arm or a leg, blind or deaf — who would be waiting a long time to come home, just because some Americans wouldn't buy enough bonds. His appeal made five sales, the newspaper reported.

Jack also got a taste for politics in Prescott. As one of the few Republicans in Yavapai County, Bobby could count on personally meeting the GOP candidates looking for local support, including a two-time candidate for governor, rancher Bruce Brockett. Arizonans voted heavily Democratic in those days, and Brockett got a drubbing — losing 40-60 percent — in both 1946 and 1948.

Jack got to see political power in action through his uncle, Ed Locklear, a heavyweight in the Democratic Party. Uncle Ed was close to the two men who had defeated Brockett: Gov. Sidney P. Osborn, who had won four two-year terms before dying in office in 1948, and his successor, Dan Garvey. Jack's family ate dinner regularly with the Locklears, and he remembered how his uncle would get calls from the governor for advice on appointments and hot issues. Years later, it would be Jack who was advising governors, both Republican and Democrat.

A young, pretty, and outgoing widow, Bobby would have had no trouble finding dates. She wasn't interested, though. Not while the boys were still growing up. But the lonely nights must have been hard. Bobby was drinking, and it would become a problem that she couldn't overcome. Jack didn't say it outright, but in later life, he indicated obliquely that he regretted never finding a way to intervene.

Meanwhile, Jack thought he'd found a career direction in school. An algebra teacher had gotten him hooked on math. It started a lifelong passion for education: "He kind of reached in my brain and ignited some embers which, really to this day have not been fully extinguished."[5] When Jack got to high school, he concentrated on science and math. His senior year, he won the Bausch and Lomb award for the student with the highest grades in math and science.

Jack's ambition when he graduated from high school in 1951 was to be a petroleum engineer. But the road was blocked. The University of Arizona, in Tucson, was the state's only full-fledged university at the time, and it didn't have a program in that field. Jack couldn't afford to go to an out-of-state school. So he settled on metallurgical engineering, which a counselor suggested as the closest alternative.

Jack was lucky to be able to go to college at all. It would proba-bly have been impossible without the insurance policy his father had taken out so many years before. He worked summers and vacations, sometimes at two jobs, but he needed that initial financial foundation. He saw it as a kind of scholarship, and he was committed later on to making sure that other students got the same opportunity.

"The insurance policy was just the little bit of help I needed; it was the motivation to get me through (school)," Jack said in an inter-view. "I simply would not have been able to go to school without it."[6]

Jack had rarely gone beyond Prescott except for a visit to the Grand Canyon and periodic trips to Phoenix — including a junior high adventure to the Arizona State Fair, when he and a friend traveled by bus and stayed at the YMCA. Still, he said, "I thought I was a very sophisticated senior high school student." The feeling evaporated the minute he got to college. When his mother drove him to UofA and dropped him at the dormitory, it was the first time he'd ever been in Tucson. "I was totally bewildered," he recalled.[7]

UofA was a whole new world for the small-town boy. The number of students, 5,588, was nearly enough to populate Prescott. The university town was booming. The city had 45,000 residents in the 1950 census and would hit nearly 213,000 by 1960. (In that same period, Prescott's population didn't quite double, going from 6,764 to 12,861.)

Mack Wiltcher was already at UofA when Jack arrived. Neither of them took school very seriously at first. They joined a fraternity, Theta Chi. "We weren't very good students," Mack said, "because we were having too much fun, partying. The first thing we would do when we signed up for classes was to determine how many legal cuts we could have."

They all liked beer and would go on "boondockers," parties out in the desert with a healthy supply of alcohol. Or they'd sit around for hours drinking coffee and talking. Jack played pool, and he made money at it.

"Jack was so darned smart, he didn't have to study too hard," Mack said. And Jack knew how to organize his time. If he sat down for an hour of studying, he focused for the full time.

Still, he failed freshman English. The professor, Jack remembered, couldn't stand engineering majors and took every chance to humiliate them.

The professor had a habit of going around the room every week, asking students what they were reading. When Jack kept answering, "nothing," the professor took to calling him the "class anti-intellectual." But the snide label had a happy consequence. It shamed Jack into joining the Book of the Month Club. The first to arrive in the mail was *The Caine Mutiny*, Herman Wouk's Pulitzer Prize-winning novel about the ethical fallout of decisions made at sea by a Navy ship's captain and crew.

Jack opened it up when he got home from school in the afternoon, couldn't put it down, and read straight through until he reached the last page the next morning. "That really started me on a lifelong career of reading," he said.

Jack lost a party buddy when Mack dropped out after his second year: His father had a heart attack and Mack went to work in the mines. He got engated and asked Jack to be best man. But when a priest somehow heard about it, Jack was forbidden, because the ceremony wasn't Catholic. The warning failed: Jack stood up with his old friend at the wedding. With the "bastard" label from his childhood and a class in world religions that opened his eyes to the varieties of faith, Jack broke with Catholicism. He was deeply spiritual his entire life, but had no interest in joining any church.

Mack and Scotty's Wedding
L to R: Bud Rolstad, Bill Esser, Stan Waitman, Jack Pfister,
Mack Wiltcher, Leola (Scotty) Scott, Shirley (Scott) Harvey,
Judy Roher, Jerry Coppinger, Dorthy Head

As Mack's family got back on its feet, he enrolled at Northern Arizona University for a while. But he was struggling with a pre-engineering course. It was just before Christmas. So he called Jack, who was coming to Prescott for the holiday break. On New Year's Eve, Jack gave him a tutoring session. They took a break for a quick champagne toast and went right back to work until nearly dawn.

Jack's summer jobs gave him a lifelong respect for manual labor. Between his freshman and sophomore years, he and Tad headed to Southern California. They worked in the oil fields during the day and on the Long Beach docks at night, loading diesel drums going to Korea. They didn't get off until midnight, but the money was good.

By the last two years of college, Jack was doing better because the workload concentrated on his field of study, engineering. Except he was starting to realize that he didn't actually want to be an engineer. But with no idea what the alternative might be, he kept plugging away at his original major. At least it would lead to a job.

And he got one with Shell Oil Co. after graduation. The petroleum giant hired engineering majors of any type and then trained them in whatever area it needed. Jack was sent to Southern California as an engineer in training.

An engineer in training

He moved into an apartment in the Belmont Shore area of Long Beach. Handily, there was an attractive woman in the apartment opposite his. Patricia R. Emerson was a year older than Jack. She was born in Sioux Falls, S.D., and after moving around, her family settled in Cedar Rapids, Iowa. Pat trained as a nurse there, got a job at a local hospital, and bought a car. Then a friend suggested a road trip.

The young woman's aunt lived in Long Beach — did Pat want to drive to California?

"I said, 'Sure, why not?'" Pat recalled.[8] But once they arrived, the friend decided to move in with her aunt and left Pat in the lurch. Fortunately, nurses were in demand. Pat found a job in a post-operative surgical ward and joined a co-worker in renting an apartment in a building filled with other young people. Jack chatted her up while she was using the washing machine in the laundry room one day. She could see his potential right away: "He was a sharp guy. He was pretty stable."

For Jack, she stood out as "the only one who had any sense" among the flock of women in the building. "I was a nurse and very practical," Pat explained. "Had to be." The two hit it off and started dating.

A student deferment had originally kept Jack out of the Korean War. Once he was working, Uncle Sam came calling again. When Jack went for his draft physical, though, his thick glasses and weak eyesight kept him a civilian. He was classified 4F: not acceptable for serving in the armed forces.

Without military service hanging over his head, Jack felt he could get married. In 1956, he and Pat drove to Cedar Rapids for the wedding and headed to Tucson for their honeymoon. Like so many things in Jack's life, choosing Tucson was intensely practical. He was about to start law school at the University of Arizona there.

Jack and Pat's wedding, Cedar Rapids, 1956

One reason was geography. The Southern California lifestyle never suited Jack. He wanted to be back in Arizona. But with a degree in metallurgical engineering, he was bound to end up in one of the state's isolated rough-and-tumble mining towns: Ajo, Douglas, or Bisbee. As a lawyer, he could live in Arizona without necessarily being stuck in the sticks.

Jack had also gotten a taste for law. His prominent uncle, of course, was a role model. But he'd also had a closer look from an unexpected vantage point. When Jack was fourteen, his uncle got him a summer job as a relief janitor at the Yavapai County courthouse in Prescott. The building might have looked like part of a Norman Rockwell painting. But the job of cleaning the bathrooms, according to his brother, Tad, was tough, smelly and, if local drunks had happened through, outright disgusting.

On the other hand, there wasn't really that much work. The first few days, Jack came down after three hours to say he'd finished everything he'd been asked to do. You can't do that, he was told. He had to stay until quitting time even if he ran out of things to do. Out of pure boredom, he started reading casebooks and files in the county attorney's office. Besides picking up some legal procedures, "I learned a lot about some of the people in Yavapai County," Jack drily observed.

Law school was a better fit for Jack. This time he really studied. One reason, he said, was being married. "I've said that it's amazing how much time I had to study when I quit having to chase girls!" he said.

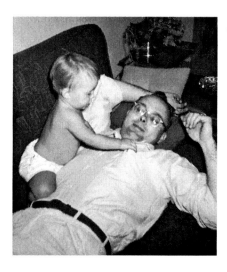

Jack with baby Suzanne

Jack had one more reason to take his studies seriously in 1957, when he and Pat had a baby girl. They named her Suzanne.

Three years later, when they had moved to Phoenix, the Pfisters had a son. They named him Scott Emerson, making Pat's maiden name his middle name.

Besides law school, Jack was working from twenty to forty hours a week as a research metallurgist for the U.S. Bureau of Mines. His performance review there is a revealing snapshot of his strengths. While "quality of work" and "work attitude" were rated "satisfactory," his work habits were judged "excellent." As for work volume, he was "outstanding."

With Pat working as a nurse at the Thomas-Davis Clinic in Tucson, the young couple had enough money to live modestly.

Along with Tad and Mack (who was back in town to finish his undergraduate degree at UofA), Jack used to go fishing in Mexico, in Libertad, a Yaqui fishing village at the time, where they had a beach house. They mostly did surf-casting. "Jack was a great cook," Mack remembered — perhaps because the other two men didn't want to cook. "He made the best fish stew you'd ever eat."

Jack finished law school in 1959 at age twenty-five, one of three students who graduated with distinction. His class rank was second or third, "depending on how you count," he said. The graduation itself was a blur for Jack, with the Arizona bar exam just a month away. He nailed it.

Jack's mother, Bobby, created a lot of opportunities for her two sons to succeed.

This was a doubly proud time for Bobby. A week after attending the law-school ceremonies in Tucson, she went to Tad's graduation from the College of Veterinary Medicine at Colorado State University at Fort Collins.

Jack hoped to become a law professor. But once again, money blocked the way. He needed a graduate degree, and he couldn't get one without a scholarship. Unfortunately, the number-one student in the class (who also one-upped Jack as editor-in-chief of the new law review) applied to the same major law schools as Jack. And for the limited scholarship money that was available, he got the offers.

Some prestigious schools accepted Jack — Michigan and Columbia, as he recalled in 1991 — but he couldn't afford them. In the end, in what must have been an ironic twist for Jack, his rival ended up refusing all the scholarships and instead went to clerk for a justice at the U.S. Supreme Court.

The Road to Salt River Project

Jack never wanted to end up in Phoenix. He and Pat loved Tucson and hoped to stay there after he graduated from law school. But the economy was still limping after the 1958 recession. Jack's job search came up empty in Tucson. His next choice was Prescott, the town that would always have a special place in his heart. But his Uncle Ed talked him out of it. Prescott was a small town and already had too many lawyers. The place to look for work was Phoenix.

Jack not only had an impressive resume, but he also had a connection. A law school classmate had landed a position at Jennings, Strouss, Salmon & Trask, and the firm was still hiring.

1959 photo of the newest lawyer employed at Jennings Strouss.

Jack clicked with Jennings Strouss from the day he started work in 1959. It was one of the larger law firms in Phoenix, yet with the addition of Jack, it had just seventeen lawyers. The timing was perfect for an ambitious, new attorney. The firm's business was outstripping its staff, so Jack immediately got responsibilities that normally would have gone to more seasoned lawyers. The senior partners would bring in files and walk off after giving little more direction than, "Here, do this." Sometimes, Jack admitted, "I didn't know what the hell I was doing."[1]

It was the era when talented women with legal minds ended up as secretaries, not lawyers. They often knew the law nearly as well as their bosses did. With help from the secretary of one of the founding partners, Jack said, "we kind of groped our way through."

One of Jennings Strouss's major clients was Salt River Project. It's a water and electric utility with a unique public-private structure. In fact, SRP, as it is commonly called, consists of two separate organizations that work together.

One is the Salt River Valley Water Users' Association, a private corporation founded in 1903 by farmers who pledged their land as collateral for federal loans to build Roosevelt Dam. The association now oversees a string of dams and an extensive canal system that delivers water to urban and rural users in the Phoenix metro area.

The other is the Salt River Project Agricultural Improvement and Power District, set up in 1937 to provide electricity to users in SRP's service area. The district was organized as a political subdivision of the state so it could issue municipal bonds. SRP thus has access to cheaper financing than private corporations. That's a sore point with Arizona's corporate producers of electricity. So is SRP's exemption from oversight by the Arizona Corporation Commission, which regulates other power companies.

Although each half of SRP has its own board of directors, practically speaking, the two end up represented by the same slate of ten members (the power association has an additional four at-large members). The voting system is as unique as SRP: The votes are based on the number of acres a person owns within the SRP service area. Farmers, of course, have the vast majority of votes and have continued to dominate the board even as SRP's territory has become more and more urban.

In 1962, the partner handling SRP, Ted Riggins, came to Jack and asked, "You're a mining engineer, aren't you?"

"No, I'm a metallurgical engineer," Jack answered.

"What's the difference?" Riggins wondered. Jack tried to explain that these were two vastly different specialties. But Riggins wasn't listening.

"What do you know about coal?" he asked.

"Not very much," Jack said.

"Well, you know more than anybody else in the firm," Riggins concluded.

The next day, Jack was on a plane to Colorado, where SRP was trying to line up fuel for a new power plant. In the past, the utility had relied on hydropower, oil, and natural gas to run generators. Now it was looking at adding coal to the mix.

Jack's companion on that first plane ride to Colorado was Les Alexander, an associate general manager at SRP who became one of the major mentors in his life. Jack described Les as a good engineer, but an even better politician. Les was a dapper dresser and a schmoozer, who loved a good time and fine wine.

A lot of Jack's legal work for SRP in the early years involved right-of-way and condemnation proceedings.

The first time he discussed a case with the board of directors, it was a disaster. SRP had lost a condemnation case in court and was forced to pay a high price for some raw desert — far too much, in Jack's opinion. The directors, however, took the court's ruling as a sign that SRP was paying too little when it bought right-of-way on farmland. They wanted to recalculate the value of the past purchases of right-of-way and pay the farmers more.

Jack had to tell them no. "I told them that legally they couldn't do that," Jack recalled. "That once they had settled, they couldn't go back and give these people more money."

Board President Victor Corbell was so furious he tried to get Jack fired. It didn't happen, but Jack didn't appear at a board meeting for several years. In the meantime, his legal work gave him a chance to get to know some of the individual board members and build good will.

Arizona was like the Greek mythic figure Tantalus, condemned to eternal thirst as he stood in a pool of water he could never reach. The state had rights to a share of the Colorado River, which runs through the north-central part of the state and along its western border with Nevada and California. But Arizona couldn't get the water. Not where it was needed. The fields of central and southern Arizona and its booming cities were hundreds of miles away from the reservoirs on the lower Colorado River.

The state had been trying for years to get Congress to approve and finance an immense aqueduct system known as the Central Arizona Project. Arizona Senators Carl Hayden and Ernest McFarland first introduced a bill to build the project in 1947. But the large and powerful California delegation kept blocking it in the House.

Negotiations on Central Arizona Project legislation were heating up in the 1960s. The major people above Jack, SRP's Les Alexander and Jennings Strouss's Ted Riggins, began spending a lot of time in Washington. That opened up extra SRP work for Jack, including lobbying.

At the same time, the Arizona legislative process was changing in ways that fit Jack's skills. The Capitol had traditionally been a cozy place where legislators would hole up and play cards with lobbyists from powerful business interests, taking the occasional break for committee meetings, recalled Jack DeBolske, who was a young lobbyist for the League of Arizona Cities and Towns.[2]

In the beginning, Jack spent a lot of time with DeBolske sitting in hallways, while favored insiders worked the smoke-filled rooms.

SRP used a lobbyist, Tom Sullivan, with a flair for wheeling and dealing behind closed doors. But after the 1966 election, the law-making process moved into the public eye. Now the supporters and opponents of bills were expected to appear at committee meetings and make presentations. That wasn't Sullivan's strength.

SRP's general manager, Rod McMullin, asked for the bright, young lawyer from Jennings Strouss to do the job. Now Jack was working closely with SRP's top executive. And McMullin began the mentoring that would lead Jack to the top.

For the next decade, Jack spent hours in McMullin's office talking about how SRP worked. Jack later described the sessions as like "one of the Hopi elders taking someone down in a kiva [sacred gathering spot] and beginning to give them all of the secrets of the organization, and all the secrets of how they did business."

His career had taken off. But one big opportunity for Jack fell through: the chance to go to Washington. After Richard Nixon was elected president in 1968, the transition team included Howard Allen of Southern California Edison Co. He'd worked with Jack on utility issues for years. Now he called to ask if Jack was interested in a post in the Interior Department — assistant secretary for water and power. The job was such a sure thing, he said, "You can just start packing your bags."

Executives in other investor-owned utilities, though, balked at putting someone from a public utility into the job. So Allen had to add more candidates. By then, Jack had gotten excited about the opportunity and was building his own base of support, with help from McMullin and Riggins.

But politics was working against Jack.

Phoenix's rival power providers had their own allies in Congress. SRP was lined up with Democratic Rep. Mo Udall, while APS was close to Republican Rep. John Rhodes. Jack met with Rhodes, who would have a strong voice in choosing the assistant secretary. But Jack was later convinced that "APS really blackballed me" with the congressman.

Jack knew he didn't have a shot when he met with the new Secretary of Interior. Wally Hickel was dealing with an oil spill in California and had no idea who Jack was when he showed up for the interview.

To top it off, Jack got caught in the bitter rivalry among the seven states that share Colorado River water. He was from the Lower Basin: Arizona, Nevada, and California. Another candidate was from the Upper Basin: Colorado, New Mexico, Utah, and Wyoming. With each group opposing a candidate from the other side, they canceled each other out. The new assistant secretary was from Missouri.

SRP reorganized in 1969. Les Alexander got responsibility for the power side. He wanted Jack to leave Jennings Strouss and come to SRP as his special assistant.

The offer was enticing except for one thing: It meant a big salary cut. Jack was torn. One day he wanted to make the move. The next day he didn't. He had a great career going at Jennings Strouss, where he could pick and choose his work. The firm had put him on a fast track, making him a partner less than a year after he started.

He got into a rare argument with his brother, Tad, over the offer. "Why in the world would you ever give up a month's vacation, the salary you're making, the benefits you have, to go to SRP?" Tad asked.[3]

On the other hand, Jack could see that his job at Jennings Strouss was bound to change. SRP had just set up its own law and contracts departments, so much of the work he enjoyed would now be done in-house. He was reaching the ten-year mark at Jennings Strouss, which brought some financial benefits that would ease the pinch of earning some 20 percent less than before.

He decided yes. Before going to work at SRP in January 1970, he sat the family down to explain. Suzanne was in eighth grade when her father told them, "We're not going to have as much money. We're not going to be able to do as much. But I'm going to be much happier."[4]

In truth, Jack was never much interested in the financial trappings of power. Neither was Pat. "People had this assumption that we grew up lavishly, because they knew him when he was an executive," Suzanne said.

The family always shopped at mainstream, economical stores. "We were J.C. Penney kids," she recalled. "We lived simply growing up."

The Pfisters initially lived in a modest, mass-produced house in a development called Maryvale. Jack's hobby was woodworking, which was a handy skill in the early years of his marriage. He made several pieces of furniture when he and Pat couldn't afford to buy anything new. He built a gas station for Scott and a dollhouse for Suzanne.

In 1966, the family moved to a larger house in a new, middle-class subdivision in north-central Phoenix. While it was under construction, Jack amused himself by building an architectural model of it.

The one-story, cement-block ranch house had 2,724 square feet, after an addition that increased its size by about a third in 1984. "It's not a grandiose house," Suzanne said. "It's not what you'd think an executive's house would look like."

Her parents never showed the least inclination to move, however. Jack was too comfortable, and Pat had her fill of being uprooted after a childhood in which her family moved nine times in twelve years.

As Jack's career progressed, "He could have lived anywhere he wanted, in any kind of mansion," said former Mayor Terry Goddard, "and he continued to live in this frumpy little house."[5]

His home reflected Jack's approach to issues. "Part of his genius, I thought, was that he sort of mastered the art of being unprepossessing," Goddard said. "His soft voice, his demeanor — which I think was by design. He could say the most extraordinary things in a most gentle way."

Coal and Clean Air

The Central Arizona Project finally received Congressional approval while Jack was still working at Jennings Strouss. President Lyndon Johnson signed the Colorado River Basin Project Act, which included the vital canal, on September 30, 1968. But there were strings attached that would keep Jack busy for years.

One was the question of power. The Central Arizona Project would need an enormous amount of energy to pump water to Phoenix. The original plan called for a pair of immense, hydroelectric dams, which would pump water through the system and also help pay for the project through sales of excess electricity. But they would flood two scenic canyons on the Colorado River, and the Sierra Club launched a national campaign to kill the dam proposal.

"Should we also flood the Sistine Chapel…?" one ad famously wondered.

Congress listened. When it approved the Central Arizona Project, the dams were out, replaced by a requirement to find an alternative source of power by September 30, 1969 — or lose federal funding.

The stakes weren't just water for Arizona, but also electricity for the Southwest, where growth rates for utilities were running 10–15 percent a year. A committee with representatives from nine utilities — with Jack among the SRP contingent — began working on a proposal for a coal-fired plant. The plan was to expand an existing plant. But at the last minute, one California company pulled out, and the whole deal fell apart.

Jack's boss, Les Alexander, saved the day. He pulled together SRP staff working on the project, and they sketched out a proposal for a power plant on the Navajo Reservation. The location, just east of the small town of Page, Arizona, was near two essential ingredients for generating electricity: water and coal. Salt River Project would manage the construction and operation. The other utilities were skeptical. But nothing else was on the table, and they agreed to SRP's proposal for the Navajo Generating Station. They just met the deadline.

The Navajo Nation signed a lease agreement on September 29, 1969. The construction cost would be $309 million for the plant and $171.6 million for transmission lines. The government would pick up 24.3 percent of the tab. SRP had the next largest share, at 21.7 percent, and four other utilities, including Arizona Public Service and Tucson Gas and Electric Co. (now Tucson Electric Power), would pay the rest.

When Jack came to SRP at the start of 1970, he was put in charge of coordinating all aspects of the Navajo Generating Station. He also handled issues for SRP's power side, which Alexander headed.

The plant was a massive undertaking. Its three units were scheduled to be completed one after the other in 1974, 1975, and 1976. The trio of 775-foot smokestacks would be among the tallest structures in Arizona. The coal would come from Black Mesa on the Navajo and Hopi reservations and travel seventy-eight miles by rail to reach the plant.

Jack had to pay attention to a lot of little things. That included donuts, coffee, and hundreds of bags of flour. You couldn't hold a meeting on the Navajo Reservation, he learned, without offering some hospitality and gifts. People needed a reason to make the long drives to the chapter houses, the tribal administrative buildings.

Jack and his SRP colleague Leroy Michael hashed out agreements on rights of way for transmission and rail lines. They negotiated how to handle preferential employment on the project for Navajos. They put together the coal-supply contract, even though "neither of us had ever written one before," Michael said.[1]

The plant's April 1970 groundbreaking was just a month away when pollution threatened to become an ugly fight. Arizona was in the midst of setting emissions standards for power plants. Just twenty miles from the north rim of the Grand Canyon, the Navajo Generating Station was in a particularly sensitive spot. The state Board of Health proposed a thirteen-ton daily limit on coal ash released from the plant. That meant preventing 99.4 percent of the ash from escaping into the air.

But in March, SRP and Arizona's two other major utilities asked for the standard to be set at nineteen tons — nearly 50 percent higher.

When the Board of Health held a hearing in May, the battle lines were drawn. Conservation groups came armed with arguments to defend tighter pollution controls. Then Jack spoke. And they were startled to hear that SRP and its partners had done a U-turn. The utilities urged the board to adopt the originally proposed thirteen-ton standard.

Working behind the scenes with state health officials, Jack had negotiated a compromise that made the lower limit palatable: The Board of Health would give the plant some flexibility by making allowances for breakdowns and looking at twenty-four-hour average emissions, which would allow for short spikes in pollution.

The deal showed what would become the hallmarks of Jack's style: Work outside the glare of publicity; focus on practical steps to bridge differences; make sure no one feels like a loser.

◇◇◇

SRP, as was typical in the industry, had shrugged off environmental questions in the past. When he started work there in 1970, Jack recounted in an oral history, "they had no environmental department. They probably couldn't even spell ecosystem. And it was just not something that was of importance to them."[3]

But now SRP had to pay attention.

The groundbreaking National Environmental Policy Act (NEPA) had taken effect on January 1, 1970. The law required an environmental analysis of any major project that involved federal funding, permits, or government-performed work.

What about Navajo Generating Station? It was planned before NEPA passed, but constructed afterwards. Was it grandfathered in or did it have to meet the new requirements? There was no clear answer.

For environmentalists, however, here was an opportunity to take on America's insatiable appetite for energy. Their prime target was pollution from coal-burning plants. Jack traveled around the Southwest debating environmentalists about power generation and defending Navajo Generating Station. He initially wrote off his debate opponents as a bunch of radical hippies. As he listened to them and read more, however, he came away convinced that there was a lot of merit to what they were saying. SRP clearly needed to re-examine its strategy and catch up with the times.

Jack could look to his boss as a role model. Les Alexander had good instincts and picked up on the importance of environmental issues. With McMullin's approval, he changed the Navajo Generating Station's design to include pollution-control devices that raised overall costs by nearly a third.

Jack recommended creating a separate environmental department. Both Alexander and McMullin agreed — and they put Jack in charge of organizing it. He argued successfully for making it a free-standing group that didn't report to any other department. "I thought it needed to be independent," he explained later, "and to be kind of the conscience of the Salt River Project."

In 1969, Jack heard from a young engineer working in southern California. Al Qöyawayma, a Hopi from the reservation's Third Mesa, had written about possible job openings.

Jack didn't make a job offer. But he remembered the letter when he ran into Qöyawayma the next year at a Flagstaff forum on the value of coal-fired generation and its impact on tribes of the Colorado Plateau. Qöyawayma spoke about some of the technical and cultural issues.

Afterwards, he recalled, "Jack came up and introduced himself in his affable and calm way."[2] Jack and Pat were heading off to visit the Hopi Reservation, and they invited Qöyawayma to join them for dinner at their hotel.

The purpose was more than social. Jack wanted to sound out Qöyawayma about the job of heading the environmental department, where the main focus would be the Navajo Generating Station.

Qöyawayma had impressive credentials. With degrees from California Polytechnic State University and the University of Southern California, he had gone on to develop and patent inertial guidance systems and star trackers for Litton Systems.

Jack saw a perfect fit in other ways. "He needed somebody sensitive to native affairs, but who wasn't an activist and had technical knowledge," Qöyawayma said.

With business out of the way, Qöyawayma took Jack and Pat to see his aunt Polingaysi, an author and artist. She had taught her nephew pot-making, and Qöyawayma would later become a nationally recognized potter and sculptor.

When Qöyawayma arrived to take the job at SRP in 1970, Jack wheeled over an entire cart piled high with material about the Navajo Generating Station. With a wry smile, he said, "Here, Al, this is all yours." It wasn't just a quip, it was a signal of Jack's management style: Assign responsibilities, set expectations, and step back. No micromanaging.

Over the years, SRP's environment department's work ranged from tracing air currents with a smoke generator on a biplane to setting up a program to track bald eagles. At one point, a farmer complained that nearby transmission lines were making his cattle sick. SRP ran a study, grazing cows in a field near transmission lines, then putting them down and conducting autopsies to check for adverse effects. There weren't any. Meanwhile, the public wondered what would happen to the cows that were killed. "We're going to serve them in our cafeteria," was Jack's joking response, according to Qöyawayma.

Jack was sympathetic to tribal interests. "He didn't wear native culture on his sleeve," Qöyawayma said, "but he was sensitive to native people as much as to his own board." When the Hopi held a training session and retreat in southern Arizona, Jack came down and patiently answered questions. He went out to dinner several times with Hopi Tribal Chairman Vernon Masayesva.

Do you know any Native American engineers? Qöyawayma got the question again and again as SRP job recruiters came up empty. The answer was: Virtually none. Finally, he and six other Native American scientists decided to make an organized push. In 1977, they founded the American Indian Science and Engineering Society.

Qöyawayma asked Jack for some start-up help. The response was simple: "Do it." Qöyawayma was able to take the time he needed, tap secretarial staff, and get material published. For years, SRP bought an ad on the last page of the society's quarterly magazine. AISES now has high school, college, and professional programs and has awarded scholarships to nearly 5,000 students. "The seed, the necessary nurturing from a practical standpoint, was provided by Jack," Qöyawayma said.

Jack's first effort to comply with the National Environmental Policy Act was a "Through the Looking Glass" experience. The U.S. Interior Department was arguing *against* doing an environmental impact statement for the Navajo Generating Station. SRP, in a role reversal, was asking *for* one.

Jack was looking several moves ahead. Skipping a formal, environmental process might be easier in the short run, but it could lay the foundation for legal challenges in the future. When SRP insisted, the Interior Department did a superficial analysis that Jack considered virtually worthless. Fortunately, he'd become friends with an official in the Interior Department who shared his views about the importance of creating solid environmental impact statements. The two men quietly rewrote the documents for the Navajo Generating Station. In the end, Jack said, "They were of sufficient quality that [they] really precluded environmentalists from filing lawsuits."

With his growing base of knowledge, Jack quickly became involved in environmental issues at a bigger scale. In August 1970, he and an executive at Arizona Public Service were named to a newly formed Environmental Policy Committee that brought together a wide spectrum of interests, including environmentalists and scientists, to advise Western electric utilities on siting plants and other facilities.

The goal was to address environmental concerns at the initial planning stages of power installations.

For the Navajo Generating Station, however, the environmental battles were far from over. A growing haze of pollution had blurred the views at many of America's iconic national parks, including the Grand Canyon. In 1977, amendments to the Clean Air Act required improved visibility in national parks and wilderness areas, with strategies that included limiting emissions from large fossil-fueled power plants. The owners of the Navajo Generating Station, including SRP, insisted that the plant played only a minor role in the Grand Canyon's haze.

Environmental groups, on the other hand, accused federal regulators of failing to enforce visibility requirements for national parks. They sued in 1982, and five years later, the National Park Service did a test tracking the Navajo Generating Station's emissions of sulfur dioxide, a gas that reacts with sunlight and other chemicals to form particles. Drawing on the test results, the Environmental Protection Agency in 1989 identified the plant as a major source of haze at the canyon. The agency's next step was to decide what anti-pollution technology to require, which might be sulfur scrubbers costing $200 million to $1 billion.

Jack responded with a lengthy opinion piece in the *Arizona Republic* on September 10, 1989, disputing the test methodology and interpretation. Other studies, he argued, identified the major culprits as smelter operations, auto pollution from Southern California, and natural phenomena. SRP believed, he wrote, that "sulfur-scrubbing equipment at Navajo Generating Station will not make a difference in what can be seen at the Grand Canyon." The plant owners were willing to prove their point by sponsoring another emissions study.

But the weight of evidence pointed to Navajo Generating Station. Jack was no longer at SRP in 1991, when the plant owners agreed to install scrubbers. The debate wasn't over, though, and it flared up two decades later. In the kind of consensus building Jack would have appreciated, representatives from a variety of interests, including environmentalists, tribes and SRP, worked out a plan: One of the plant's three units would close in 2019 — or emissions would otherwise by reduced by a third — and additional pollution controls would be installed by 2030. On the industry side, some objected that the cuts went too far; on the environmental side, some argued that the pace was too slow. But the EPA adopted the proposal in 2014 as its final regional haze rule.

Taming Wildcats

Help! Ten-year-old Scott Pfister used a stick to write the word in the sand. He'd tagged along with his father, uncle, and four other men on a fishing trip to Mexico over Thanksgiving weekend in 1970. They were camping on a remote beach twenty miles from Puerto Libertad, and now one of their cars had broken down.

The man in the suit also loved to get a little messy at times, especially when he was fishing.

Jack and his brother, Tad, made the trip to Mexico, a day's drive from Phoenix, at least once a year. The trips involved "lots of drinking, swearing, and great food," Scott remembered. "Dad would spend a few weeks of what little spare time he had to prepare for these trips."[1]

Jack would pour lead weights, put his cook box together, and make sure the Coleman stoves and lanterns were working. Scott suspected his dad enjoyed the preparations more than the trip itself.

The dead car was a very temporary adventure. When the men drove into Libertad in the other car, they found a vacationing telephone worker who happened to be a part-time mechanic. It was the middle of the night, but he and a buddy went with Jack and Tad and got the car running.

Jack wasn't someone to let good deeds go unrecognized. He wrote a letter to the vice president of Mountain Bell praising this "most outstanding example of service" from the two employees.[2]

That year, when Jack started at Salt River Project, brought sadness, too. His mother, Bobby, died of a stroke.

Jack's career was accelerating. In 1971, an executive shuffle opened up the position of director of operation services. Jack got the job, reporting directly to general manager Rod McMullin. Jack's domain included land purchases, warehousing, and transportation.

Before, the only person Jack had supervised was his secretary. Now he managed an entire staff. And he had no background in business.

So Jack did what he always did when he had a new challenge. He started reading. "He was a voracious reader," his daughter, Suzanne, said. "He was a lifelong learner before it was fashionable."[3]

Jack bought stacks of management books and holed up with them every night. He embarked on "a sort of self-taught MBA," Suzanne said. He remained fascinated with management theory and practice throughout his life.

Over the years, several of the mainstream standards became his guideposts. One of his favorites was Stephen Covey's *The 7 Habits of Highly Effective People*. It articulated much of what Jack had already learned to do: Begin with the end in mind, look for win-win, really listen to others, and find strength in teamwork.

Right away, he focused on achieving a participative style of management. Seeing supervisors who were rigid and autocratic, Jack began systematically swapping them out.

Some of the changes required a ruthlessness that didn't come naturally to Jack. But he was determined to put the right managers in place. Bob Mason, who worked in the water area, said Jack explained it this way: "Sometimes you have to do things you don't want to do. Sometimes when you stick the knife in, you have to turn it."[4]

Jack's growing responsibilities began squeezing out the time for fun, like hosting bridge parties and what Scott called "awesome cocktail parties." Still, he managed to fit some woodworking into his schedule and make toy guitars for Scott and several friends, who "played" and sang along with records in the garage.

Within a year, Jack was moving up again. Les Alexander retired in the fall of 1973, and Jack took over as associate general manager for power.

There were doubters. Jack's spectacular rise left some people, especially those who were passed over, questioning how competent he was. Others were skeptical about a leader who didn't know the power business. Once again, Jack started a self-education program, including traveling around to visit SRP sites.

When linemen staged a wildcat strike in 1974, Jack got pulled into labor relations. The whole staff — some 200 linemen — plus around 550 sympathizers walked off the job on Monday, March 4. Media reports said scheduling issues led to the strike, which wasn't sanctioned by the International Brotherhood of Electrical Workers. But Stanley Lubin, the attorney hired by the strikers, said the real issue was the way Salt River Project was sitting on grievances instead of arbitrating them.[5]

Negotiations became a peculiar dance. They were held in an upstairs conference room at the Caravan Inn, a now-demolished motel just east of downtown Phoenix. Lubin wasn't allowed into the room, since the strike was unauthorized, so he spent hours in the bar downstairs.

As the week dragged on, a federal mediator was brought in, and Jack joined the upstairs negotiating team. The mediator would pass messages between Jack and Lubin.

The sticking point was how the strikers would be treated when they returned to work. SRP wanted to fire some and impose lengthy suspensions on others. The strikers wouldn't accept a deal that cost anyone's job.

SRP twice set, and then delayed, a return-to-work deadline. On Sunday March 10, though, general manager McMullin issued a "last ultimatum": If the strikers didn't go back to their jobs Monday morning, they'd all be fired. And not only would they lose their jobs, they would forfeit their pension benefits.

That day, Jack pulled Lubin aside to talk. The board was going nuts over the continued walkout, Jack said, and he offered a scaled-back set of sanctions, although it still included some firings. When Lubin agreed to take it to the strikers, Jack commented that he knew where they were meeting. You've had someone following me? Lubin asked. Yes, Jack answered.

The strikers shouted down the proposal. So Lubin played his trump card: He had resignation letters from all the strikers. If they left on their own, they'd still be eligible for their pensions, a multi-million-dollar cost for SRP.

He sent the mediator up to Jack with a message: If a settlement couldn't be reached by the end of the day, Lubin would submit the resignation letters. Fifteen minutes later, Jack came down alone. Whatever anger he felt, he didn't unleash it on Lubin. "He knew that I wasn't on strike," Lubin said. "He also understood the fact that my clients are entitled to an attorney."

Jack agreed to limit the sanctions to suspensions. The scheduling disagreement would be submitted to arbitration. Jack recognized that the strikers had a point about prolonged delays on grievances, Lubin said, and he agreed to limit postponements to sixty days.

A little later, Lubin, Jack, and SRP's attorney on the walkout went to lunch together.

Looking back on Jack's role in the strike settlement, Lubin said, "I've always admired the way he handled that."

In 1982, the men dealt with another wildcat strike at the Navajo Generating Station. This time, SRP hit the union with a legal demand for damages, calculated to the penny. Local IBEW officials opposed the walkout, and Lubin rushed to end it. Salt River agreed to withdraw the lawsuit and not to fire anyone. Lubin got a call from Jack afterwards. "He congratulated me on settling it," Lubin said. "He appreciated the fact that I understood the way it works: When you don't have the cards in your hands, you fold."

Although Jack had little interest in the social whirl of the well-heeled, he decided in 1974 that Suzanne should go through a formal debut topped off with a cotillion. He told the organizer that his goal was to get his seventeen-year-old daughter to stop saying "bullshit." Another factor may have been that an SRP employee and his wife had just started the Arizona Honors Debut. It was designed to be more about merit than money. "We learned how to pick china, but we also met with women doctors and lawyers," Suzanne said. "They really drummed into us that this was a different kind of cotillion."

The ball, though, had all the trappings of a classic coming-out event, with the young women in long dresses and white gloves. Jack did his share of drinking, and by the end of the night he was acting a little silly. It was the last time Suzanne remembers seeing her father intoxicated.

Later, he gave up alcohol altogether. Although there's no sign that Jack's drinking was out of hand, he worried about alcoholism in the family. His own mother had suffered from a drinking problem later in life.

Several times Jack arranged interventions with friends and colleagues who had alcohol problems. One time he was leaving Phoenix Country Club with a friend who had been drinking heavily. As they waited for their cars, Jack told him, "I'm worried about your drinking."

"That hit me like a sledgehammer," the man remembered. Soon afterwards, an attorney approached him about going to Alcoholics

Anonymous and helped him enroll in a program at St. Luke's. Jack certainly had a hand in the encounter, which ended up being the turning point toward sobriety.

But the tactic wouldn't have worked without that initial shock from someone as universally respected as Jack. "If he had not said something to me, I would not have listened to the attorney," the man said. "It really got my attention, a lot more than anything else ever did."[6]

◇ ◇ ◇

Jack was disciplined about nearly everything. Except his weight. "He knew for his health it was better to be twenty to thirty pounds thinner," Suzanne said. But it was one goal that eluded him.

Jack's work day began early and didn't end when he returned home from the office. After dinner, he would spend a little time reading the afternoon newspaper and then head into his home office. He kept it relatively organized, with a folder in his file cabinet for everything, even if it had only one sheet in it. He'd come out around 10 p.m. to catch the news on TV.

Evening was also a time for calls to SRP staff. "Jack was a telephone hog," said Leroy Michael, who worked with him for decades.[7] Michael's family knew to set aside time around the dinner hour every night for the Pfister call.

But despite the pressure of work, Jack made time for parenting. He remembered his experience in athletics as "a good character-building exercise," and he wanted his son to have similar opportunities. So he coached Scott's Little League team. It was "a very unrewarding experience," he said years later, "because the parents get very angry at the coach when their children don't play."

To help teach Scott to pitch and sharpen his control, Jack created a batting zone: "It was this amazing little pitching device," Scott recalled. "A pole and straps and eye hooks."

Although Jack had never ice skated, he and his next-door neighbor coached in a hockey league. "All we did was to make certain that the right number of kids got out on the rink," Jack said. "They didn't need much coaching; they were just kind of skating and hitting."

Jack joined Scott in the YMCA's father-son program known then as Indian Guides. He was scoutmaster for Scott's troop. Pat, meanwhile, was heavily involved in Girl Scouts with Suzanne.

Jack was more involved with his children than many men in his generation, "but probably not as involved as I would like to have been." When he talked to retired people, he would later tell university students, "they always say they regret not having spent more time with their children. And I'm included in that category."[8]

At the Top of the Ladder

Jack didn't expect to rocket to the top of Salt River Project in just six years. He wanted to move up the corporate ladder, but when he came to SRP in 1970, "it was not with the idea in mind that I would become the General Manager."[1]

Looking back, it seems clear that Jack was being groomed by Rod McMullin to be his successor, maybe even while Jack was still at Jennings Strouss. By November 1975, Jack had been promoted to deputy general manager, the second-highest executive position. And yet as McMullin prepared to retire in 1976, everyone was wondering: Just who would he recommend to become the next general manager?

"Rod really played all of that very close to the vest," Jack said. He told the board that he had two candidates: Jack and Bob Amos. He was still evaluating them and would make a recommendation later.

Changing leadership was an unfamiliar process. McMullin had been general manager for nineteen years, seemingly as much a part of SRP as its dams and canals. He was also the first overall chief executive SRP had ever had. Before that, there was a separate top executive for each of the two organizations that make up the utility: the water users' association and the agricultural improvement and power district.

The suspense was driving the board members crazy. They even resorted to plying McMullin with booze at an out-of-town conference in spring 1976, but they still couldn't pry out the name.

Finally, in May, McMullin named Jack as his choice. The board members were twitchy. Jack was just forty-two, too young and inexperienced, they worried, for such big responsibilities. They even suggested that McMullin stay on, but he insisted Jack was ready.

In the end, the board approved McMullin's choice. Jack became general manager on July 1.

Jack Pfister became general manager at SRP on July 1, 1976.

Despite all of McMullin's coaching, there was a lot Jack wasn't prepared for. One was the high visibility. Jack had spoken at public forums, he'd been quoted in the newspaper, but now he was the public face of SRP. When things went wrong, Jack would be the target of anger and frustration. He would stand for Salt River Project in editorial cartoons. Customers would blame him personally when anything went wrong. Within weeks of taking the top job, he was criticized for a "philosophy of inevitable rate increases" in a letter to the editor in the state's biggest newspaper, the *Arizona Republic*.[2]

"I did not really understand the pressures that were imposed on the General Manager to become a quasi-public person," Jack said. "Up until that time I had been a fairly private person."

He hadn't expected any problems with the board. But now he had to assert his authority.

The board was entrenched and insular. The members came from the farming interests that had founded the Salt River Project. They were elected by district in low-key elections with little turnout or publicity. Once in office, they virtually never lost and stayed on for as long as a half-century or more. Their major concern was keeping their water rates low. Power sales subsidized water, and they were determined to keep the electric side profitable.

The rural makeup of the board was at odds with an area that was rapidly urbanizing. When Jack became head of the Salt River Project, barely half of the land within its service boundaries was still agricultural. But he would struggle to get the board to recognize how SRP was changing, let alone deal with the issues of its urban service area. SRP's city customers used relatively little water, and their focus was on the electric bills.

"I had a lot of self-doubts about whether I was really up to handling the job," Jack admitted later.

He knew SRP would have to keep raising electric rates — although that would certainly set off cries to rein in its independence and regulate it the same way as investor-owned utilities. The oil embargo of 1973-74 had ignited a spate of inflation, which didn't let up until the early 1980s. SRP's costs were soaring: a flagman's vest that went for $2.90 in 1974 was $5.61 in 1976. It also needed to invest $2 billion over the next five years to handle rising demand.

SRP had gained ten thousand customers in 1975, raising the total to 249,000, and it expected to reach 342,700 by 1981. Serving them all would take more generating, transmission, and distribution facilities.

Months after becoming general manager, Jack urged the board to approve a rate hike that would increase electric bills by an eye-popping 21.3 percent (which, based on new financial information, he later revised upward to 24.9 percent). Since then, "I have been running from angry electric customers," he joked in a personal letter.

That seemed high enough. But SRP had other pressures. It had become a partner in a string of power plants, including the yet-to-be-completed Palo Verde Nuclear Generating Station. The projects were financed by bonds that required certain levels of revenues. The Salt River Project had to keep Wall Street happy to keep raising money for construction projects, said Richard Silverman, an attorney for the utility who later became general manager.[3]

Jack got a lesson in board politics. Board members were more worried about SRP's ability to sell bonds than a backlash from consumers. They had hired a consulting firm that advised a whopping rate boost of nearly one-third. Thanks to board President Karl Abel, who became an ally of Jack's, the amount was ratcheted down to 26.7 percent. It was the fifth rise in rates since 1972.

"We will have a difficult time explaining to customers why the rates are going up the way they are going up," Jack said. "No one is going to be happy…"[4]

Jack brought a very different style to SRP leadership. McMullin was comfortable in a cowboy hat and could play the good ol' boy with board members who wore Levis to their meetings. Jack was more formal. He always wore a tie at the office and, according to his daughter, owned barely one pair of jeans. (When he was unwinding, he wore khakis and a button-down shirt.)

He made decisions deliberately, building an extensive base of knowledge first and getting input from all angles. With his intellectual curiosity and wide-ranging interests, "He was an egghead dropped in with a bunch of farmers," said Bill Shover, retired director of public affairs at Phoenix Newspapers Inc., which published the *Arizona Republic* and now-defunct *Phoenix Gazette*.[5]

McMullin, who had risen to the rank of major in World War II, saw his job through a military lens. He was a top-down boss, both autocratic and mercurial.

Jack didn't want a rigid chain-of-command structure. He wanted to know what was going on — which led to all those nighttime phone calls. But from his reading of management theory, he knew the importance of delegating responsibility. Once a project was clear, Jack expected those below him to proceed on their own: "He'd give you enough rope to hang yourself," one observed wryly.[6]

Jack took a personal interest in employees. "He had a remarkable ability to remember names," said Bill Davis, who worked in environmental services at SRP.[7] At a large company luncheon one day, he greeted some thirty people on his way to sit down. He not only knew them all by name, Davis said, but he asked about the projects each was working on.

As a CEO, "he was just all around the best," said Sid Wilson, who reported to Jack and later headed the Central Arizona Project.[8]

Jack explained his success to Wilson this way: "I never had very many original great ideas, but I had this great ability to recognize other people's great ideas and then pull people together to put that great idea into play."

Jack was already busy with industry and civic organizations. He was on the board of directors of the American Public Power Association, the Metropolitan YMCA and the Girl Scouts.

In his personal as well as his corporate life, Jack was always "trying to cram 16 hours of activity into 12 hours time," the company newsletter, *Pulse*, noted in a July 1976 piece about the new general manager.

"I'm very demanding on myself, and, as a result, I'm very demanding on the people who work for me," he said in the interview.

Jack somehow jammed even more projects into his schedule. He was one of just two local executives who served with Phoenix officials on a subcommittee to reorganize city government.

SRP's structure was in the spotlight, too. The voting system for the board was based on the amount of land a person owned in SRP boundaries: the more acres, the more votes. So while the number of urban customers was exploding, farmers continued to have all the muscle. Jack was acutely conscious that SRP's one-acre one-vote system was increasingly outdated, although it survived a U.S. Supreme Court challenge.

The Legislature rebalanced the equation somewhat in 1976. It expanded the board of directors on SRP's electric side, adding four at-large members, elected on the basis of one vote per landowner. That would give more clout to urban customers, few of whom owned even an acre. Voting for the remaining ten board seats would still be weighted by the number of acres owned.

SRP got Jack into political circles. Suzanne remembers coming home from college for a visit in 1976. Jack was going to a big fundraiser for Rep. Mo Udall and invited his daughter to come along. The Arizona Democrat was trying to raise money to pay off debts from his unsuccessful bid for the presidential nomination. "I had never seen my dad work a room like I saw that night," Suzanne said. "I remember being agog at how he went up and talked to everybody and schmoozed. I thought he was a pretty subdued guy, and he was out talking to everybody and laughing and jovial."[9]

Jack also got SRP into politics. In 1976, Arizonans were voting on a ballot measure, Proposition 200, that would have effectively pulled the plug on nuclear power by requiring legislative approval for any plan to build a plant. At the same time, Jack could see that government was stepping up regulation of pollution, water, land use, and other areas with a direct impact on SRP. The utility couldn't simply tend to its canals and dams. It had an enormous stake in who got elected to office.

SRP set up a political action committee in mid-1976. "We had a responsibility to get involved," Jack explained. "Just last year, more than 70 bills which would have had an effect on some aspect of our operations or our customers' interests were introduced in the state Legislature alone."[10]

Jack and SRP board president Karl Abel suggested which candidates the committee should support. They chose both Democrats and Republicans, with a criterion of backing "people that generally support the overall philosophy of the free enterprise system."

Going into the election, SRP itself planned to put $174,600 into defeating Proposition 200. Jack called it a proper corporate expense "to protect our interests."[11] Employee volunteers campaigned against the measure. To Jack's relief, as an advocate of nuclear power, voters turned it down.

Critics objected that SRP, as a political subdivision of the state, should not be allowed to jump into political activities. But with the project's unique structure, that identity applied only to the agricultural improvement and power district. The water users association was a corporation — and Jack made sure that the political action committee was strictly tied to that half of SRP.

Jack and Pat did their part to help candidates. In a personal letter, he painted a droll picture of how he and Pat, as political novices, hosted coffees for the GOP candidates in their district.[12]

The first one was "a disaster," he wrote. They had inadvertently invited the worst possible mix of people to listen sympathetically to the three Republicans' pitch: a "woman's libber," two environmentalists, a social worker, a libertarian and three Democrats. One candidate then droned on so long that the other two had virtually no time to speak. Still, the trio managed to make statements that offended everyone in the room.

In a post mortem, they agreed to limit the talks to five minutes, "during which the candidates would talk in glittering generalities." If anyone asked a tough question, Jack and Pat would announce that it was time for refreshments.

"As chief coffee scheduler," Jack wrote, "I developed an uncanny ability to schedule coffees on nights when we could be sure that only the most dedicated and committed Republicans would show up."

Although Jack dealt with deeply serious subjects his whole life, anyone who knew him never forgot his laugh. It was genuine, hearty, and inclusive. It broke the tension when meetings were getting testy. His puckish sense of humor showed in the tongue-in-cheek start to a speech in Flagstaff at a western regional Farm Bureau conference in August 1976. The Federal Energy Administration had just issued a convoluted statement about a pickup in energy imports and the easing of the current economic slowdown.

"In order to clarify the cautious terminology of the experts," Jack said, "it should be noted that a slowing up of the slowdown is not as good as an upturn in the down curve. It is hard to tell, before the slowdown is completed, whether a particular pickup is going to be fast. At any rate, the climate is right for a pickup this season, especially if you are about 25, unmarried, and driving a red convertible!" [13]

SRP under Jack became a pacesetter in conservation, with a major program to help residential customers use less electricity. "Power saver advisors" did free home inspections, offering advice on ways to save energy. SRP sold and financed attic insulation. While such programs are common now, SRP's Power Saver Service was the first of its kind in Arizona and one of the few in the nation at the time.

"We believe we have a responsibility to our customers to help them conserve energy and keep electric bills as low as possible," Jack said.[14] On the other hand, as a newspaper article pointed out, SRP wasn't changing its rate structure, which charged less for industrial and other customers who used the most electricity.[15]

When Jack moved into the general manager's chair, he also took McMullin's place on a powerful, year-old civic group. The Phoenix 40 was launched in 1975 by newspaper publisher Eugene Pulliam, heavy-hitting attorney Frank Snell, and Tom Chauncey, owner of radio and TV stations. They invited prominent local men to join them in tackling the problems that were holding back the state's largest city.

Traffic was a mess in a sprawling metropolis with virtually no freeways and little mass transit. Schools and higher education had lackluster supporter. Arizona was the land-fraud capital of the country, where raw desert was peddled to thousands of unwitting Americans as developed land.

In many ways, it was vintage Old West politics — a group of unelected people taking solutions into their own hands. The Phoenix 40 saw itself in a "fight against community apathy and indifference," according to a news article marking its first anniversary.[16]

At the end of a year, the group's accomplishments included establishing task forces on crime and transportation, looking at alleged problems in the county attorney's office, lobbying for a statewide grand jury, and endorsing Arizona's bid (ultimately unsuccessful) to land a federal solar energy research institute.

Jack was the facilitator at the Phoenix 40's first retreat, said Dennis Mitchem, a retired partner at Arthur Andersen and a member of the original Phoenix 40.[17] Mitchem, who was president of the Chamber of Commerce at the time, had seen plenty of professional facilitators in action. "Jack was far superior," he said.

Members arrived at the retreat in a dither over the term "Phoenix 40." Did it sound too elitist? Should they change it? Jack refused to let the conversation get sidetracked: "The reason our organization exists is to deal with the problems of the state, not agonize over what our name is," he said.

Jack made sure everyone had a chance to talk, but he wasn't afraid to lead. When the subject of education came up, one person said the topic was too big. The Phoenix 40 should leave it alone, he warned, because they could never get their arms around it. Jack quickly countered, "It's also too important for us to forget about it."

One big question for the group was privacy. Some members didn't even want the roster to be public. The meetings were supposedly closed. But after one particularly scrappy gathering, someone talked to a reporter. The next day's newspaper had a story all about it. With the inevitable prospect of publicity, Jack argued that the Phoenix 40 should open its meetings to the public.

There was a nuance in the way Jack operated that could seem contradictory. He wasn't categorically opposed to closed-door meetings on public matters. He accepted that at some point negotiations on controversial issues may have to be private and one-on-one. But he believed in keeping the decision process as above-board and transparent as possible.

Reporters knew that it was useless to ask Jack to go off the record. On the upside, they could always quote him. On the downside, they never got any insider secrets.

When Jack became chairman of the Phoenix 40 at the start of 1978, he set a quiet course. In an interview seven months later, he made it clear that the group wasn't going to throw its weight around. "We just haven't found all that many instances in which we feel there is a role for Phoenix 40 to play," he said.[18]

The community was already tackling major problems. Plus, the Phoenix and state chambers of commerce had taken a more hands-on, effective approach.

"It's only when we think we can add something that isn't being added that we get involved," he said.

Boom! Work on the Central Arizona Project began with an actual blast on May 6, 1973. Arizona Gov. Jack Williams and Interior Secretary Rogers C.B. Morton detonated a charge at Lake Havasu. The lake, actually a reservoir behind Parker Dam on the Colorado River, was the starting point for the massive aqueduct system that would extend to Phoenix and then Tucson.

The controversy over the Central Arizona Project was far from over, however. Jack would be pulled in again and again to help resolve issues before Colorado River water actually flowed out of taps in Arizona.

Opponents with a variety of objections had organized in a nonprofit group called Citizens Concerned About the Project. Their issues included cost, loss of wildlife habitat, and the impact on Indian reservations.

The biggest fight was over dams. While Congress had stripped the controversial Colorado River dams out of the Central Arizona Project, it still authorized a major new dam in Arizona.

Orme Dam was designed to store Colorado River water and provide flood control, with secondary benefits that included generating electricity. A mile long and 195 feet high, it would be built at the confluence of the Salt and Verde rivers, twenty-five miles northeast of downtown Phoenix. The cost of building the dam and reservoir complex was put at $223 million in 1975. The reservoir would destroy long green belts of riverside vegetation, a desert oasis that supported an astonishing variety of wildlife. Biologists found 157 species of birds at the site,[19] the highest count in the nation. Bald eagles, which had been pushed to the brink of extinction, nested there.

Orme Dam would have submerged two-thirds of the Fort McDowell Yavapai Community reservation, which spreads across forty square miles northeast of Phoenix. Some 300 tribal members lived there full time, fiercely attached to their land. In 1976, they rejected the federal government's offer to buy their land, pay compensation and relocate them.

All the while, construction was moving along on the Central Arizona Project. It was one-quarter finished when there was stunning news. The president wanted to kill it. Jimmy Carter had campaigned on using tax dollars efficiently, cutting government waste and practicing conservation, especially in energy. In February 1977, he issued a budget that proposed axing nineteen ongoing federal water projects as economically unsound and environmentally unsafe. It was quickly dubbed the "hit list," and the biggest hit of all was the Central Arizona Project.

The objections to the Central Arizona Project included the amount of power needed for pumping, water loss from evaporation, water quality, the huge price tag, the wisdom of encouraging development in the desert, and the impact of Orme Dam.

President Carter said the administration would review all federal water projects before issuing a decision in April on funding cuts for the 1978 budget.

Congress was in an uproar. Bringing home federal dollars for dams, reservoirs, and canals was a time-honored tradition.

In Arizona, Gov. Raul Castro quickly appointed a task force to get funding for the project back into the next year's budget. The group was heavy on business leaders and included heads of the two big electric utilities, Jack and Keith Turley, president of Arizona Public Service.

When the Interior Department held hearings on March 21, the task force rushed to Washington to defend the Central Arizona Project.[20] The session lasted nine hours. Governor Castro warned that if the aqueduct weren't built, "a catastrophic day of reckoning lies ahead" because the state was pumping groundwater twice as fast as it was naturally replenished. He pointed out that the funding wasn't a gift but a federal loan that would be completely repaid over fifty years.

Jack testified about the need for water to generate electricity for Arizona's growing population and economy. Without the Central Arizona Project, utilities would have to build power stations along the Colorado River, with all the environmental risks of those locations and the massive transmission lines needed to reach urban areas.

Sen. Barry Goldwater explained that the project's importance had been well documented for decades. He made an emotional plea to the panel: "Forget the economics of the project and consider whether my grandchildren will be able to live in Arizona."

Not everyone was rooting for the project. The heads of five Arizona tribes testified as neutral, since they favored some parts of the project and opposed others. And almost thirty foes turned out to wrap up the hearing.

Fortunately for Arizona, Interior Secretary Cecil Andrus was a Westerner. He was sympathetic to arguments that the state couldn't tap its share of the Colorado River without the giant aqueduct. He tried to persuade Carter to limit the hit list to projects that were "real dogs," with harmful impacts and negligible benefits for their high price tags.[21]

When the budget was announced in April, members of the task force could sigh with relief. The Central Arizona Project survived, albeit with a budget trim. But Carter had killed Orme Dam. The state would have to come up with an alternative way to store water. And it faced a new condition for the funding of the Central Arizona Project: Get control of its groundwater pumping. They were both issues in which Jack would play a major role.

Pfister's Air Force

Labor problems were brewing at Salt River Project in 1978. The utility's contracts with the International Brotherhood of Electrical Workers were expiring at the end of the year. A union election turned ugly, with personal attacks on individual SRP executives. The friction between labor and management had been growing since the 1974 wildcat strike, when some union members were fired over their behavior on the picket lines.

Local 266 of the IBEW represented some 2,500 SRP employees, including clerical staff and the hourly workers who operated its power plants. When Rod McMullin was general manager, SRP had taken a tough line with the union and its business manager, Don Hall. Now Jack faced a union leader itching to fight. He realized Hall had decided that "the time had come for labor to really show Salt River Project who was boss."[1]

Jack hoped to head off a showdown. "It's been my experience in labor negotiations that talks are most successful when they deal with the issues in an atmosphere of mutual respect, with courteous and tactful behavior," he wrote in a company newsletter.[2]

SRP's working conditions, wages, and benefits, he argued, were very competitive with the utility industry in the region and with other employers in the community.

But management and the IBEW couldn't reach a deal before the contract expired on December 31. SRP made a final offer with 6.5–9 percent pay raises. Union members overwhelmingly rejected it in early January.

The two sides continued to talk, including three sessions with a federal mediator. On January 10, though, they reached an impasse.

That day, as union workers got ready to hit the picket lines, SRP put out a special edition of the company newsletter, *Info*. Jack urged calm and was already laying the foundation for a return to normal.

Nonunion employees should be courteous to picketers and avoid getting pulled into arguments, he advised. "After the walk-out, many strikers will want to come back to work. We will all need to work together to heal the wounds and to begin to build SRP again."

At the same time, Jack made it clear that management wasn't going to play softball. Employees who went on strike could be replaced, he warned. And they would lose their health and life insurance benefits the minute they walked out.

The strike began at 2 p.m. January 10. It was the first union-sanctioned walkout against SRP in a quarter century. Jack reacted by immediately putting SRP's final offer into effect, showing workers what they had to gain.

Pay wasn't the only issue, however. The main points of contention were SRP's plans to consolidate two field offices, trim the number of employees working on holidays, contract out custodial work, reduce the extra pay for temporary supervisors at the Navajo Generating Station, and take some people out of the union by making them salaried rather than hourly workers.

Hall fumed over the special status of the power side of SRP's business, the Salt River Power District. As a quasi-municipality, it didn't fall under the jurisdiction of the National Labor Relations Board. "Thus, they do not have to answer to the government or anybody else for what they do in the way of labor relations," he complained.[3]

Both sides were under pressure. The union provided no strike benefits to members. On the corporate side, the walkout forced SRP to stop new electric hookups, close three of its five business offices, and temporarily suspend residential irrigation service.

Supervisors and salaried workers suddenly found themselves out in the field, learning to do new jobs. A purchasing agent drove a forklift, unloading spools of wire. The associate general manager for water put on boots and a hardhat to work on irrigation structures in the mud. For many, the working day stretched to ten or twelve hours.

The struggle focused on the Navajo Generating Station. The union put it "under siege," Jack said.

Strikers blocked the way when SRP sent other workers to run the power plant. The first employees who tried to cross the picket line were harassed so badly that the utility started using buses to take them into the plant. The union insisted on boarding every bus to make sure that no hourly employees were aboard — "a very humiliating thing" Jack said, "but we really had no other alternative."

The second day of the strike, windows of two buses were broken and their tires were slashed. SRP said the incident was an attack; Hall called it an accident. Although no one was hurt, violence was now another sore point between union and management.

Someone cut an overhead electric wire for a coal train near Page — it had no impact on power production, since SRP had built up a sixty-day supply of coal in case of a strike. But Jack blamed the strikers for the vandalism. He complained that an employee was run off the road and several had received intimidating phone calls.

Jack had refused a union proposal to go to binding arbitration, which he thought would abdicate management responsibilities. Now he said SRP wouldn't negotiate until the union stopped all acts of violence and terrorism.

Hall countered that SRP or some other party "is creating this violence to make us look bad," and as a pretext for calling in the National Guard or federal forces.[4]

As the walkout continued, Hall put the blame on Jack personally: "Mr. Pfister just wants us to strike for some reason, and the reason is to break the local. We have tried to get back to the bargaining table with a federal mediator and he refused to go."[5]

"It really became a test between Don Hall and myself," Jack said in his 1991 oral history.

The union leader called Jack "a power hungry man trying to force his way on the people."[6]

There were signs attacking Jack. An anonymous note circulating at SRP portrayed Jack as a puppeteer manipulating negotiations. A special edition of the corporate publication *Info* invited hourly workers to return to the job and laid out SRP's explanation of why talks broke down. A worker used it to scrawl a note to Jack at the bottom of the page that began: "Mr. Pfister — You go to hell for lying…"

Meanwhile, the strike was taking on the trappings of a spy novel. There was evidence that hourly employees had tapped telephone lines so they could listen to conversations between plant management and the Phoenix office. So SRP bought scramblers for the phone system.

SRP was using a guard agency and asked it to infiltrate the strikers. To Jack's surprise, the undercover agents turned out to be three young women. "That was, they told us, the most effective way and these were experienced women who knew how to handle themselves in this kind of a situation," Jack recalled in his 1991 oral history. "We began to get very reliable and useful information about what the strikers had planned."

The state Department of Public Safety had its own spies in the union camp. Jack met regularly with a senior DPS officer to compare notes and work on strategies. Several times, Jack recounted, there were very serious threats against some of the power-plant employees or their families. DPS, he said, was "very helpful in making certain that the families were going to be protected and in calming me down a little bit."

Strikers had been locking arms to keep out supply trucks at Navajo Generating Station. To get around the blockade, SRP brought in a Huey helicopter and flew materials to the power plant. The local newspaper in Page called it "Pfister's Air Force." Later, an executive group gave Jack a flight cap emblazoned "Pfister's Angels," playing off the name of a popular TV detective show, *Charlie's Angels*.

SRP went to court to open up the supply route. It got restraining orders from Maricopa County Superior Court and, after some delay, the Navajo Tribal Court, ordering union members not to interfere with deliveries to the generating station. But enforcement was difficult. The director of the Office of Navajo Labor Relations argued that the court had no authority to get involved.

It wasn't until January 23, nearly two weeks into the strike, that four supply trucks finally got into the plant, thanks to a police escort that included DPS and the Coconino County Sheriff's Office.

Now that suppliers had access, SRP was willing to consider going back to the bargaining table. Negotiations resumed on January 27. A federal mediator ran marathon sessions that lasted three and a half days. When they finally wrapped up on Monday, January 30, there was a tentative deal.

That night, Jack was in Page, along with several of his top executives, to meet with local residents at the request of a leader of the Church of Jesus Christ of Latter-day Saints. Harvey Gardner was alarmed at how the strike was tearing friendships apart. A mother at the gathering told how her daughter had lost her best friend, because one girl's father was salaried, while the other was hourly.

Jack pledged to the group that no union member would be blackballed. But he cautioned, "I am not going to tell you we can wipe the slate clean," according to the February 1 *Page Paragraph*. "It's going to be difficult. The wounds do not heal immediately. It will take constructive attitudes on management's part and on the union's part."

On February 1-2, union members voted on the proposed contract. It passed. On the dollars and cents side, the contract differed little from SRP's final offer. Hall, though, claimed victory based on some SRP concessions, including fifty-four hours' notice for employees not working holidays and retraining or job protection for custodians whose work was contracted out.

The road to the final deal wasn't quite as straightforward as it was reported. As often happened when Jack was working on an issue, there was a side conversation that advanced the solution. In the middle of the strike, an international representative for IBEW, Raymond Duke, was in town and had breakfast with Jack. He saw the weakness of Hall's position, according to Jack. The two men outlined the broad terms of a settlement.

Jack also recognized that his own negotiating team had been inflexible, probably trying to prove how tough they could be. So he put a new person in charge, Leroy Michael, "without a doubt, the best negotiator that Salt River Project had." The changed dynamics were key in wrapping up negotiations.

The strike had lasted twenty-three days. SRP followed up with newspaper ads that had Jack's trademark appeal for unity. In the *Lake Powell Chronicle* and the *Page Paragraph,* newspapers serving the area around the Navajo Generating Station, an open letter published February 8 called for "a new foundation built on mutual understanding."

"Strikes are constructive," it stated, "if they result in greater appreciation of the job by workers and of the worker by management, reduce smoldering discontent and provide for a fresh start."

Deluge in the Desert

The Salt River's normal flow through the Phoenix metro area is zero. The water has been captured for decades in reservoirs behind a string of dams managed by Salt River Project. The last of four dams to be built on the Salt — and the farthest downstream — is Stewart Mountain Dam, completed in 1930.

The Verde River flows into the Salt a few miles below Stewart Mountain Dam. SRP intercepts the Verde with two dams, Bartlett finished in 1939 and Horseshoe added upstream seven years later.

The final link in the SRP string is Granite Reef Diversion Dam, twenty-two miles east of downtown Phoenix, which channels water from the rivers into a canal system.

Arid Arizona, which includes a vast swath of the Sonoran Desert, has two seasons of rainfall: Winter brings long, steady rains while summer has thunderstorms, which can be violent and highly localized. But the amount is unpredictable, and the state tends to swing between long, dry periods broken by strings of wet years. As a result, the rivers can be trickles or torrents.

When farmers originally banded together to form Salt River Project, their biggest worry was having too little water, not too much. They built dams to store water for their fields and, as an added economic benefit, to generate electricity — not to protect the surrounding urban area. "The system of dams we currently operate is not designed to provide adequate flood control for the Valley," Jack explained in an April 1,1979, *Arizona Republic* Q&A.

There's a key structural difference between flood-control and storage dams. Flood control requires huge outlets near the bottom, so large amounts of water can be released to free up space for incoming water. Storage dams have relatively small valves to release water for farmers, municipalities, and other users. Their spillway gates are close to the top, and large quantities of water can only be released when the reservoir is full.

Phoenix boomed when the Salt River below the Granite Reef dam was just a desiccated ribbon of dirt and gravel. From 1941 to 1965, as the population grew from 65,000 to more than 500,000, the river never flowed through the metro area because the reservoirs on the SRP system were never full. Maricopa County, which takes in virtually the entire Phoenix metro area, didn't even establish a flood-control office until 1959. Planners and developers blithely subdivided the riverbed and the flood plain next to it, and houses, stores, and businesses went up as if the river would never flow again. Bridges over the dry riverbed weren't designed to withstand flood-stage flows, while many river crossings were simply roads at grade.

While Phoenix and surrounding cities shrugged off the risk of floods, they were actually in the bull's eye. The watershed of the Salt and Verde rivers is larger than the state of Maryland — and it all drains toward the Phoenix metro area.

For its part, SRP "had operated on the assumption that it had no flood-control responsibility at all," Jack said.[1]

The river wouldn't stay dry forever. In late December 1965, heavy rains and runoff from melting snow in the mountains north and east of Phoenix were filling reservoirs dangerously close to capacity. An entire generation of SRP dam operators never had to worry about releasing water from its dams. Now they scrambled to figure out how to do it. Once they did, water came rushing down the Salt River through Phoenix. Some 8,000 people were briefly evacuated as a precaution.

Five years later, in early September 1970, central Arizona got a drenching from the remnants of a Mexican hurricane that dumped some thirteen inches of rain on Mount Ord, north of Roosevelt Lake, and four to five inches of rain in parts of Phoenix and Scottsdale within a twenty-four-hour period. It was a "100-year storm" — so severe that it's expected to occur just once a century. The flooding didn't come just from the river, but from SRP's canal system.

The Arizona Canal, built in 1883, is an open, earthen-banked canal that runs roughly east to west through the northern part of Scottsdale and Phoenix. Heavy rains had caused several breaks in the Arizona Canal over the years. Two sections collapsed in 1970, flooding more than two hundred houses in Scottsdale.

The homeowners sued and won. SRP appealed to the state Supreme Court. In 1976, the justices upheld actual damages but dismissed punitive damages. What looked like a partial victory, however, was actually a big defeat: The justices rejected SRP's argument that it had no responsibility whatsoever to prevent flooding. Instead, they said SRP had to exercise reasonable control over the water in its canal system. "Ultimately, it meant we had to get into the flood-control business," Jack said.

For many at SRP, it was "a traumatic mental process," he said, and some executives resisted doing anything. But general manager Rod McMullin reluctantly began re-examining how SRP operated. When Jack took over the top job a few months later, it was becoming clear that "we had a long ways to go" in figuring out how to manage floodwaters.

So did the county and state. Storms in 1970, 1972, and 1973 triggered flooding along the Salt and other rivers. But Maricopa County voters rejected a bond issue to fund flood-control projects. In 1973, the Legislature finally limited construction in flood-prone areas. Cities and towns were required to map them and control development there. But four years later, almost none had followed through.

Then, in 1978, the deluge came again. A storm that began February 28 dumped more than a foot of rain in some parts of the watershed.

Water poured into the reservoirs along the Salt and Verde rivers. On March 1, Salt River Project had to open the spillways on the dams, sending a torrent down the Salt River.

The sight of water actually going down the Salt River was so startling that it nearly stopped the traffic on the nearby Maricopa Freeway. Gawking drivers went as slowly as 4 mph, the *Arizona Republic* reported.

The rest of Arizona was just as thunderstruck as the drivers. "The whole state, including Salt River Project, was just totally under-prepared for flooding," Jack said in his 1991 oral history. "Police departments didn't know what to do. The Department of Emergency Preparedness didn't know what to do."

There were no maps to show the impact of releasing various volumes of water from the dams: How high would the river run? Which spots were vulnerable to flooding?

The Department of Emergency Preparedness had the only pumps to get water out of flooded areas. But the department said it would need two weeks to assemble the pumps and get them working. So SRP took the pumps. "We got them prepared overnight, "Jack said. "Salt River really provided tremendous leadership in that."

Salt River Project established a makeshift emergency center, the room that had been used as the strike center, and nicknamed it "the Crisis Center." In a cramped space that "proved grossly inadequate to our needs," representatives from power and water operations were tracking reservoir levels and calculating where flooding would occur.

Cave Creek Dam had them sweating. Operated by the Maricopa County Flood Control District, it was fifteen miles from downtown Phoenix. Unlike the Salt River Project dams, this was a genuine flood-control structure, built after Cave Creek wash caused a disastrous flood in 1921. Hundreds of homes were inundated back then. The state Capitol had two feet of water on the first floor and sodden records in the basement. Now, the fifty-five-year-old structure was no longer considered adequate protection, and the U.S. Army Corps of Engineers was busy building Cave Buttes Dam, an earthen dam a mile downstream.

Would the old dam hold? The rain was coming down in torrents on Thursday, threatening to pour over the top and undermine the structure.

If the dam gave way, a wall of water would roar through suburban neighborhoods as it headed downtown. Workers were putting sandbags around the state Capitol complex, just in case, and the Legislature had adjourned until Monday.

An official from the flood-control district was taking in the bad news in Salt River Project's crisis room. He was paralyzed with panic. Opening the gates at the dam to release water would head off the risk of a catastrophe. But that would cause flooding and damage on its own. He froze at the idea.

Three times Jack asked him to authorize opening the gates. Three times he failed to respond. Finally Jack made it an order: "Direct that the gates be opened."

"So we literally took command of the flood-management system in the Maricopa County area," he said.

Salt River Project diverted the water released from Cave Creek Dam into the Arizona Canal, but the amount was far beyond its capacity. At the peak, the water from Cave Creek was 2.5 times more than the canal could carry. Plus, much of the dirt excavated for Cave Buttes Dam was swept into the canal, building a barrier downstream that pushed water into surrounding areas.

West Phoenix was hardest hit. The *Arizona Republic* reported that the floodwater from the canal was moving at twenty-five mph, inundating much of a four-square-mile area and sending up to four feet of water into houses.

Gov. Wesley Bolin declared a state of emergency on Thursday, March 2, and went up in a helicopter to see the devastation firsthand. The metro area was nearly cut in half, with the fast-moving floodwater heaving up in four-foot waves and just three bridges open over the Salt River.

City and Maricopa County planners had assumed that the proposed Orme Dam would be built, preventing large-scale floods. So they skimped and built small bridges over the Salt River at several major crossings, including Scottsdale and Hayden roads. But the savings were washed away when floodwater carried off the approaches to both bridges.

A major sewer line, which ran under the Salt River's bed, clogged up, forcing the Phoenix Waste and Sewers Department to release sewage directly into the flowing water.

The floods took a human toll.

Three men in a land cruiser were trapped in the roiling Salt River near Mesa. The Arizona National Guard was trying to rescue them when they were swept away. Two drowned and one was missing and presumed dead.

National Guard helicopters made repeated rescue runs, picking up twenty people stranded on a roof, a couple trapped in an eroding gravel pit, and a man stuck on a sandbar.

Water poured into parts of Sky Harbor International Airport, cutting a cable and knocking out radar. Incoming flights managed to continue, spaced out to allow for the missing radar. But the raging water undercut 2,500 feet of the south runway, tearing apart asphalt and carving a ten-foot hole in the taxiway. The airport estimated the total repair bill at $5 million.

On Saturday, March 4, a thunderbolt struck Arizona, but not from the storms. Governor Bolin had died of a heart attack during the night. Bolin, sixty-eight, had become governor just five months earlier when Gov. Raul Castro was named ambassador to Argentina. Under the Arizona Constitution, the secretary of state (Bolin) was next in line to become governor.

When Bolin died, the governorship went to the attorney general, thirty-nine-year-old Bruce Babbitt. In the coming years, Jack and Babbitt would work together on high-profile issues. But when he was sworn into office, the new governor had barely met one of Arizona's major leaders. In a brief encounter on the campaign trail, Babbitt saw Jack as rigid and resistant to change — an impression that he'd later find was completely off the mark.

By Monday, March 6, the rains had finally subsided and Arizona was cleaning up. It was a huge job. Almost all Salt River crossings were still closed, and two bridges had been washed away. Phoenix officials discovered late Monday that a thirty-six-inch sewer pipe had broken, pouring sewage from all of south Phoenix into the Salt

River. Unable to shut off the line, which would back sewage up into homes, or to repair it, until water levels dropped, the city resorted to dumping chlorine into the river.

As the level of damage sank in, the blame game began. Many pointed the finger at Salt River Project.

The flood had taught him, Jack said wryly, that "when the water is in the reservoirs it belongs to the people. When it is out of the reservoirs it belongs to SRP."[2]

Some complained that they didn't get enough warning about the flooding danger. In the small town of Cashion, doubly in jeopardy from its location at the confluence of the Salt and Gila rivers, residents said they weren't told to evacuate until the water was already ankle deep in their homes.

Others argued that water should have been released from the dams ahead of the storms, so there could have been a smaller, more controlled flow in the Salt River.

Dr. Robert Witzeman, president of the Maricopa County Audubon Society, blamed the flood toll on "Arizona's outmoded water policies" and Salt River Project's "agribusiness-controlled board of directors."[3]

There were calls for better infrastructure, including flood-proof bridges. For those who had never given up support for Orme Dam, the flooding was a call to action. The Phoenix City Council passed a resolution urging President Carter to reinstate Orme Dam or a reasonable alternative.

And yet, it could have been far worse. At their regular monthly meeting, which followed the March flood, SRP directors heard that twice as much water might have roared through Phoenix if the reservoirs hadn't been so low because of skimpy runoff in previous years.

Jack wasn't relieved; he worried. Now the reservoirs were close to full, with far less room for runoff. In April, Jack invited local cities, government agencies and the media to SRP's post mortem on the flood. Looking forward, he called for the disaster to be "a learning experience to get flood control for the Valley." But they might not have long to digest the lesson. Flooding along the Salt River could be even worse if another huge storm hit while the reservoirs were still brimming.[4]

And it did. The downpours, which started December 17, 1978, dumped six inches of rain on much of the 13,000 square miles in SRP's watershed. And this time there was an extra hit: The rain melted the snowpack in the mountains. Local flooding was causing destruction throughout northern and eastern Arizona.

The Phoenix area got one break: Rainfall there wasn't as heavy as in March. And this time, authorities were better prepared. But SRP still had to open the floodgates.

It was *déjà vu* all over again. There were still only three bridges over the Salt River that could withstand releases of more than 100,000 cubic feet per second. The rushing water washed away the approaches to all the others, which were built to handle no more than 25,000-30,000 cubic feet per second. Whatever repairs had been made since March were lost.

Once again, National Guard helicopters swooped in to make dramatic rescues. Part of Sky Harbor airport was flooded again, and water washed out the radar power cables again. Hundreds of homes and businesses were damaged.

Statewide, there was a heavy toll. The flooding killed nine people and destroyed millions of dollars' worth of property.

On the heels of the flooding, there was another round of finger-pointing at SRP. The call to build Orme Dam went up again.

On the Edge of Disaster

The repeated rounds of flooding in 1978 made some Arizonans wonder if they should build an ark. Jack wanted to keep them from needing one. He argued that Salt River Project had to improve its ability to predict and manage floods. Some board members and staff were dubious about making SRP proactive. But Jack recognized the organization's legal liability and the public relations disaster when entire subdivisions were inundated.

"Here we were on television with shots of the water going to the eaves of these homes and trying to say, 'Well, it's not our fault. We don't think we have any responsibility.' It just didn't ring right," he said in his 1991 oral history. SRP would have to change the way it operated. It was a fundamental shift.[1]

One critical gap was a lack of information: SRP needed a better way to gather data for making decisions about when to release water and how much. "Our whole system was pretty primitive," Jack recalled. He sent out an SRP contingent to study flood control in other parts of the country, developed closer relations with the National Weather Bureau, and hired a meteorologist to work on staff. SRP retirees living in strategic places were recruited to put rain gauges in their back yards.

The reservoirs were managed by dam tenders, who lived on site and historically had a lot of autonomy in deciding whether and how much water to release. It was a practical system, since communication to and from the deep canyons holding the dams could be difficult, especially during storms. But Jack wanted more centralized control of the reservoirs, with engineers analyzing information. So in 1978

SRP installed a state-of-the-art radio system, with backups, connecting reservoirs with the main office.

During the 1978 floods, Jack had assigned a registered engineer to each of SRP's six dams to make sure there was technical support on site. But getting them there was sometimes a challenge. Engineer Tom Sands remembered how a colleague had been helicoptered to a dam only to learn there was nowhere to land safely. The chopper had to hover a foot or two above ground in pouring rain while the engineer clambered out, clutching his sleeping bag and supplies.[2]

So Jack added landing pads at each reservoir to his to-do list. He also had SRP install both rain and river-flow gauges across the watershed, providing hydrologists with instant access to data. By February 1979, he could say in a public forum that SRP had the best equipment available to monitor conditions upstream for the risk of flooding — although he cautioned that there was no way to give two or three days' warning.[3]

SRP pushed the rest of the community, starting with the governor, to learn from the 1978 floods, as well. "We were instrumental in beginning to get people to re-examine their preparedness. It's not surprising, because it was all new," Jack said. "No one had ever done it before. It was one of those things that until you go through the experience, you can't even begin to imagine how bad it would be."

In the summer of 1979, the Los Angeles office of the U.S. Army Corps of Engineers was studying SRP dam safety and historic flooding events. Jack sent Sands to participate, not only to provide technical support but also to make sure information about Arizona was accurate.

In Los Angeles, Sands learned about a new computer program that could model how quickly reservoirs would fill, based on river-flow data and varying levels of dam releases. What a change from the manual approach, when engineers predicted reservoir levels with pencils, graph paper, calculators, and straight edges.

When Sands returned, he brought a copy of the program with him. This was all very cutting-edge, and SRP didn't have the equipment to use it. But Jack came up with $10,000 for a Texas Instruments terminal that could connect through a telephone line with a mainframe computer elsewhere and allow SRP to make predictions.

Now the federal government was raising a new worry: dam safety. The U.S. Bureau of Reclamation began reviewing the conditions of dams after the catastrophic collapse of the Teton Dam in 1976, which killed eleven people in southeastern Idaho and destroyed millions of dollars in property. The bureau issued a report warning that seventeen Western dams were likely to fail during a "maximum probable flood."[4] They included two that SRP operated, Roosevelt and Stewart Mountain.

Massive "100-year floods" were turning into an annual event. And the worst were on the way.

Storms in January 1980 were a set-up for disaster in Phoenix. They caused higher-than-average runoff and dumped snow in the mountains, leaving especially water-heavy snow at lower altitudes.

Then a pair of storms barreled across California and Arizona starting Wednesday, February 13. They caused massive flooding from the very start, closing the Santa Ana freeway in Southern California. SRP's water releases were so large they washed away ten thousand dollars in repairs that the state highway department was making at a Salt River crossing in Mesa. By Friday, the flooding was so widespread that the state Department of Public Safety asked people to minimize local travel and skip holiday trips over the upcoming Presidents Day weekend. The surging water split the metro area in half, closing off all at-grade river crossings and, by late Friday, all but two bridges.

The storm drill was familiar at SRP by now, and Jack was comfortable enough to go with Pat to a banquet of the Electric League of Arizona on Thursday. It would have been hard to skip, since the trade association, which represented more than six hundred businesses, was honoring him with its annual "Man of the Year" award.

But the break from work was short. Jack had Pat drive him straight from the dinner to SRP. He didn't come home for three days.

When he arrived at SRP headquarters, the outlook was terrifying. The "Noah's Ark" scenario for Phoenix was a series of heavy storms, coming when the reservoirs were full, the ground was saturated, and it held a heavy snowpack that would melt under the rain.

And now Stewart Mountain Dam, holding back a lake of water above Phoenix, was on the verge of failure. The gracefully thin concrete arch, 212 feet high and flanked by concrete abutments, was finished in 1930. What looked like a solid mass of concrete, however, actually was coming apart. Engineers had known for years about two big problems: Chemical reactions were weakening the concrete, and horizontal sections of concrete weren't bonded together.

They thought the dam had stabilized. But in November 1979, samples drilled out of the structure's core showed the damage was getting worse. If the reservoir water level rose too high, Stewart Mountain could have a catastrophic failure. "It's there one second and gone the next," as Sands explained. A 100-foot wall of water would hurtle toward Phoenix with nothing to stop it.

A third storm was on its way and predicted to pour an additional three inches of rain onto the watershed. Sands ran the computer simulation: With that much rain, even with all the gates wide open, the water level behind Stewart Mountain Dam would rise above the critical elevation — and the dam could immediately fail. On Friday Jack called a 6 p.m. meeting of representatives from state, county, and local governments to explain the danger. "They were just petrified," he recalled. "Nobody wanted to take charge." As Jack left the meeting, he realized he'd have to call the governor himself.

Gov. Bruce Babbitt had sent an aide, Dale Pontius, to the session and got an emergency proclamation ready. The governor had spent the day consumed with other issues: a meeting about nuclear waste and an appeals court opinion that jeopardized an already-started prison project.

He was back at his home in north-central Phoenix when a rattled Pontius called at 6:30. "Governor," he said, "we've got a very big problem with the Stewart Mountain Dam. The people at the Salt River Project think the dam may burst. It could happen tonight. It most probably will happen tomorrow when the next storm comes."[5] The danger was no news to Babbitt. He'd spent hours listening to Jack explain the problems after the previous bout of flooding in Phoenix. Jack had warned him that the current dams could not contain the monster flood that was bound to hit the Valley someday.

When Jack reached Babbitt Friday evening, he recalled telling him, "Someone's got to take charge of this problem, because if the dam breaks, we'll have chaos and we need to get prepared." Babbitt sprang into action, calling an emergency meeting at 7:30 in the National Guard Armory disaster center in Papago Park.

The Phoenix Suns were playing that night, and at least two people had to be paged out of the basketball game: the governor's chief of staff, Andrew Hurwitz, and C.A. Howlett, special assistant to Phoenix Mayor Margaret Hance.

When they got to the armory, "You could tell it was a real crisis, because there were little beads of sweat on Jack's otherwise bald head," Hurwitz remembered.[6]

Unlike the crowded make-do conference room at SRP during the 1978 flood, the National Guard operations center looked like a movie set. The walls were covered with maps. There were projections of weather and runoff data and a massive communications center. Everyone sat around an enormous table.

While Jack was visibly tense, the potential calamity weighing heavily on him, he laid out the possibilities calmly and matter-of-factly. It was almost reassuring. "You had some confidence that he really knew what he was doing," Howlett said.[7]

In fact, the meteorologists had updated the weather forecast, calling for less rain. Sands ran another simulation and saw that under the latest predicted conditions Stewart Mountain Dam should hold. But the possibility that the forecast could be revised again could not be ignored.

And Jack didn't want to downplay the danger to the public. "We felt we needed to be completely honest with people," he said. "We'd never experienced such a combination: high inflows, projections of four inches of rain across the watersheds, and nearly full reservoirs."

"If there was even the remotest chance the dam could fail, people needed to know," he said.[8]

So at a press conference, Jack explained the concerns about the safety of Stewart Mountain Dam and said engineers would monitor it over the weekend. He assured people that SRP did not believe a disaster would happen, but if it did, the Phoenix area would have forty to fifty hours' notice, the time it would take the water to reach the metro area. "There is no reason to panic at this time," he said.[9]

Babbitt's mantra all weekend was "If we err, it must be on the side of safety." He vividly recalled how Gov. Dick Thornburgh of Pennsylvania was criticized for his slow response to the Three Mile Island nuclear accident the previous March.

At the armory meeting, Babbitt told officials to prepare for the worst-case assumptions and get the word out to the public. "People need time to let things sink in," he said. "We must heighten their awareness. Overreacting is better than trying to tell them at 4 a.m. for the first time." People's first response to an emergency, he observed, isn't panic, but disbelief. And then they won't move.[10]

Babbitt went on TV late that night with a sense of urgency and danger. The Valley could see a monster flood that forced the evacuation of 200,000 people. "I don't know that it's going to happen," he said, "but I think we must be prepared. We must be prepared for the unthinkable."[11]

The river was expected to crest between midnight Friday and dawn Saturday. He urged anyone within a mile of it to stay up listening to radio or television.

At the SRP emergency center, the night dragged on and on. Jack and others closely tracked how the storm moved and changed, minute by minute, mile by mile. As daylight approached, it was clear that the storm wouldn't be as intense as they'd feared, that the heaviest rains had veered away from the Salt River watershed. The worst-case scenario wouldn't happen. But the current damage was bad enough. And there would be more as the storms kept coming over the next few days.

The *Phoenix Gazette* ran the poignant tale of a sixty-year-old woman and her sixty-eight-year-old disabled husband, who lost their retirement home to the Salt River.

On Thursday, February 21, floodwaters once again snapped a sewer line — spewing 35 million gallons a day of raw sewage into the churning Salt River.

The Valley was still cut nearly in two. But the rail bridge over the Salt at Tempe had remained open. The Arizona Department of Transportation and Amtrak, with the cooperation of Southern Pacific, hastily put together a temporary commuter train between Phoenix and Mesa. It was dubbed the Hattie B, after Babbitt's wife.

When the danger was over, Arizonans were once again dishing out blame for the flood. After a week with virtually no sleep, Jack had to defend SRP at a hastily convened legislative panel on Friday, February 22. He also had news. The SRP board had approved feasibility studies for raising the height of two dams to provide storage capacity for flood control: Roosevelt Dam on the Salt River and Horseshoe Dam on the Verde. The legislators, preoccupied with berating SRP, may not have noticed, but it was a dramatic change in policy. SRP was willing to use its dams to prevent floods.[12]

Jack made the rounds of civic and professional groups explaining why SRP wasn't to blame for the floods. In the notice for one meeting where Jack would speak, the *Phoenix Realtor* quipped, "He's damned when he dams, and damned when he doesn't."

In fact, Jack had a provocative message: Look in the mirror. All the property damage from the floods was avoidable. At a March meeting of the Phoenix Board of Realtors, he ticked off the reasons for the millions of dollars in destruction and behind every one of them was a deliberate decision.[13] The waterlogged homes and businesses had been built on known flood plains. Development had been permitted even when all the analysis and planning documents noted the high probability that a project would be subject to flooding.

Why? Jack laid it to the complacency of the long, dry period that preceded many developments, a reluctance to restrict property rights, and "the difficulty politicians have in saying 'no.'"

Meanwhile the course of the river was changed by development in the flood plain and made it more susceptible to flooding. So did sand and gravel operations in the riverbed and the dense vegetation that flourished from the discharges of a sewage treatment plant in west Phoenix. Water releases were no longer contained by the riverbanks but overflowed and cut new paths.

"Thus," he joked, "only when the Salt is dry is it a river; when it contains water, it's a flood."

Public works projects such as bridges, sewer lines, and airport runways were built assuming that Orme Dam would minimize water flows in the Salt.

None of these problems were within SRP's power to solve.

Nuclear Power on Trial

Nuclear power had a champion in Jack Pfister. Here was a way to produce low-cost energy while reducing America's dependence on imported oil.

Salt River Project chose to go with the fuel of the future in 1973, when it became a major partner with Arizona Public Service in the massive Palo Verde Nuclear Generating Station, forty miles west of downtown Phoenix. Construction on the biggest nuclear power plant in the country began in 1976.

"It appeared the technology was sound," Jack said years later. "One of the very attractive features of nuclear power was that it had little or no impact on the environment."

Then came Three Mile Island.

America's most frightening nuclear-power accident began at 4 a.m. on Wednesday, March 28, 1979. A pump failed in the cooling system of a reactor at the Three Mile Island plant near Middletown, Penn. Then a valve stuck open and an instrument panel malfunctioned. When workers finally realized there was a problem, they mistakenly took the worst steps possible.

It was the start of five days of confusion and panic. *The China Syndrome* had just opened. The movie gave a riveting portrait of the catastrophic consequences of a nuclear meltdown at a power plant. Now, what the nuclear energy industry had brushed off as improbable fiction seemed to be happening in real life.

When the reactor was finally cooled again, half the nuclear fuel had melted. In the struggle to stabilize the plant, radiation had to be released, sending a wave of fear through the region. Schools were closed, residents were urged to stay inside, and farmers were told to cover animals and feed.

By Sunday, the immediate danger was over. The physical damage was confined to the plant, where no one was injured. The containment building around the reactor had stayed intact, preventing the widespread contamination that would occur later at Chernobyl and Fukushima. Federal regulators measured low levels of radiation emissions, with negligible impact on human health (a conclusion that some critics challenged).

The accident set off a wave of anti-nuclear protests across the nation, with demonstrators chanting "Two-four-six-eight, we don't want to radiate" and "Hell no, we won't glow." California Gov. Jerry Brown called for a moratorium on nuclear energy.

In April 1979, President Jimmy Carter created a commission to look at what had gone wrong at Three Mile Island and how to prevent it from happening again. "There was really a sense at the time that that accident might spell the end to the national commitment to nuclear power," said Bruce Babbitt, then Arizona's governor, who served on the commission.[1]

The power industry rushed to form its own advisory group. The Three Mile Island Ad Hoc Nuclear Oversight Committee, set up in April, had eight electric-utility members. Jack was a natural choice. His influential role in the industry was underlined two months later, when he became president of the American Public Power Association, the national trade association of community-owned electric utilities.

This was the perfect playing field for Jack's skills in seeking practical solutions and building consensus. His low-key, unflappable manner had to be welcome when, as one utility executive said, "It was immediately obvious that the nuclear option was on trial, probably for its life."[2]

Jack started quietly meeting with Babbitt to talk about the future course of nuclear plants. With SRP's investment in Palo Verde, Babbitt noted, "obviously he had a huge stake" in the issue. And Jack knew that when change was coming, it was far smarter to help write the plans than resist and have a system imposed on you. He became a back channel between industry and the presidential commission, a way to hash out ideas. "It was really productive," Babbitt said.

What went wrong at Three Mile Island? The presidential commission and the industry committee both found a long list of failures that led to the accident.

Federal regulators spent too much time on the minutiae of licensing regulations and too little on actual safety issues. Power-plant workers were woefully unprepared for anything out of the ordinary, and control rooms were poorly designed. Utilities had failed to share information about incidents — the type of valve that stayed open at Three Mile Island had been stuck nine times before at other plants.

The nuclear power industry desperately needed a formal system for training workers, evaluating plants, and communicating problems. The industry committee came up with one. On June 28, three months after the accident, it announced the creation of the Institute of Nuclear Power Operations.

With the help of Jack's shuttle diplomacy, the presidential commission was on board. When its report came out in October 1979, the commission officially endorsed the new institute.

Jack was joint chairman of the steering committee to set up the institute, and he was on the first board of directors. "I became comfortable that if nuclear plants were properly built and properly operated, that they did not pose a serious danger to society," he said.

By December, the institute was up and running, with headquarters in Atlanta and a handful of loaned personnel. Thirty-five years later, it had a staff of four hundred and four main programs: an accredited training system, plant evaluations, incident analysis, and technical assistance.

The institute has become a model for industry self-regulation — operating not as a substitute for government, but as a way for industry to use its own knowledge base effectively. "It's still a reference point for the way things ought to be done," Babbitt said.

"In my judgment, that was Jack's creation as much as anybody's," Babbitt said. "He saw, very presciently and in a civic sense, that the industry had to get off the defensive and acknowledge that there were problems and that they ought to be doing something about it."

(The institute showed the strength of its watchdog role in 2007, when it followed up on safety violations at Palo Verde by slapping on heavy monitoring requirements.)

Jack continued to be a strong defender of nuclear power, arguing for its vital role in the U.S. economy.

"He came into the lion's den" when he spoke at the Nucleus Club, a Democratic fund-raising group, in the early 1980s, said Claire Sargent.[3] Several hundred people attended, and, like Sargent, many opposed nuclear power.

Jack laid out the plan for the Palo Verde plant and how Salt River Project would be involved. Speaking without any slides or props, he made a thoughtful, logical case for nuclear power.

"It was the most powerful thing I have ever heard," said Sargent, a staunch Democrat who later made an unsuccessful run for the U.S. Senate. "You could have heard a pin drop. Everyone said afterwards: Wasn't he magnificent?" Even though, some added, they still didn't believe in nuclear power.

◇◇◇

Much as he supported the concept, Jack became disenchanted with the practical hurdles to getting a nuclear plant built in the post-Three-Mile-Island world.

Since 1973, when SRP and Arizona Public Service decided to build Palo Verde, stepped-up regulation and environmental requirements had pushed up construction costs and slowed down progress. At the same time, inflation was surging, with prices going up more than 10 percent some years.

Both APS and SRP were feeling the financial strain. So were customers, as the utilities kept hiking rates, with a large part of the increase due to Palo Verde.

The world had changed in other ways since 1973.

That was the year Arab countries cut off oil exports to America in retaliation for U.S. military support for Israel in its war with Egypt and Syria. With Americans vowing to reduce their dependence on foreign oil and diversify the country's sources of energy, the Arab oil embargo might have seemed to be a boost for nuclear power. But America didn't need extra power. The economy was slumping, while customers were learning to reduce their energy use through conservation and greater efficiency.

When SRP decided to invest in the huge nuclear plant, demand was skyrocketing. The plans for future generating needs assumed the feverish pace of growth would never let up. But the projections weren't holding up, while the costs were far higher than predicted.

By 1980, Jack knew that it was time to rethink SRP's stake in Palo Verde. Mark Bonsall was a young financial planner at SRP who helped crunch the numbers for Palo Verde.

Bonsall presented his analysis at an internal summit of top executives. "I was scared out of my wits to walk into that meeting," he said.[4] Jack, with his curiosity and his engineer's grasp of the facts, was especially intimidating. "I knew without any question he knew a lot more than I." But Jack took the edge off. "He always did his best to make you feel at ease."

Bonsall himself became general manager in 2011, so he later knew firsthand what a risky and heavy decision Jack had faced. If SRP shed too much future generating capacity, it would be stuck buying electricity from other sources, at much higher prices, when demand picked up. If it hung in and the customers failed to materialize, SRP would be saddled with huge extra costs. Already, the spiraling price of Palo Verde was threatening SRP's bond rating and raising its overall borrowing costs. Either way, it was an enormous bet.

SRP originally owned 29.1 percent of the plant. But while Palo Verde was still under construction, Jack made the tough decision to recommend shrinking SRP's share. It sold off 5.2 percent in 1980 and, after some initial reluctance by the board, another 5.9 percent in 1982. Over the years, the sales reduced SRP's share of the nuclear plant to just 17.5 percent.

The plant's first two units went into commercial production in 1986. The third unit was finished in 1988. The final price tag was

$5.9 billion — more than double the $2.8 billion estimate when the plant was approved by the Nuclear Regulatory Commission in 1976.

SRP's cutting its interest in Palo Verde "turned out exactly the right thing to do," Bonsall said, "and it ended up being the right size when we needed it."

Reaching Out to the Pyramids

Dams had been a big part of Jack's career at Salt River Project, from an aging one on the verge of collapse to proposed ones that never got built. Now, in 1983, he was touring a very familiar structure in a completely unfamiliar place: Egypt's Aswan High Dam on the Nile River, more than 7,000 miles from home.

SRP, which had recently celebrated its eightieth birthday, was offering advice on how to use the Aswan dam to manage an irrigation system that had been in use for millennia.

When Jack joined SRP, the farmers who ran it took a very narrow view of its interests. He had to struggle at times to get the utility to look at issues beyond the clearly marked area it served in the Phoenix metro area. Now the lens was widening to include the rest of the world.

Edib "Ed" Kirdar became SRP's point person on foreign relations. He had left Iraq in 1959 to go to graduate school at Arizona State University. When a military coup plunged the country into years of political chaos, he took a job at SRP working on water projects.

Over the years, SRP became a regular stop when the U.S. State Department hosted foreign delegations interested in American irrigation techniques. Many were from the Middle East, where the United States was trying to gain influence during the Cold War. The natural contact was Ed Kirdar, with his fluent Arabic and Turkish, expertise in hydrology, and knowledge of the Middle East culture.

In the late 1970s, Kirdar recalled, he and SRP President Karl Abel, head of the board of directors, talked about whether SRP should do more than routine tours with foreign visitors. They realized, Kirdar said, "Here we have this know-how, why don't we market this?"[1]

The opportunity came through a new program in Egypt run by the U.S. Agency for International Development. A consortium, led by Colorado State University, was working to boost the well-being of small farmers tilling the ancient fields of the Nile River valley and delta. Translated into a practical strategy, that meant improving water use and efficiency — a goal tailor-made for SRP. In 1981, Egyptian and SRP officials set up the Professional Employee Exchange Program — PEEP — to share ideas and expertise through on-site visits. It kicked off in April 1982, with Kirdar taking time out from his other assignments to lead the SRP side.

Jack seemed to be on board. So Kirdar was startled to be peppered with questions about the program. Why was SRP involved? How would it benefit? What would the Egyptians gain? Who would participate? Where would they go? With an engineer's insistence on precision, Jack asked about all the tiny nuts and minuscule bolts.

"That kind of skepticism was news to me," Kirdar said. Was Jack's heart really in the project? he wondered. But Jack was simply doing his own due diligence and was in fact deeply committed.

In its initial two years, from 1982 to 1984, PEEP brought sixteen Egyptians to SRP, while eight Arizonans went to Egypt. The international development agency paid for transportation, travel costs, training material, and medical insurance.

The exchange program gave Egyptian engineers hands-on experience in planning a 24-hour-a-day modern water-delivery system. They learned techniques in construction, maintenance, weed control, and water measurement.

SRP officials, including Jack, traveled to Egypt in 1983 to evaluate the program. On the plane, Kirdar was surprised to see Jack wasn't grabbing a chance to relax and maybe get ahead of jet lag. Instead, he was busy studying, learning as much as he could about Egypt in the window of time that was available.

Jack came back clearly impressed. Catching up with Kirdar over a cup of coffee in the cafeteria one day, he said, "What you're doing is very important for ourselves, our country, and the world. I want you to have a special department and work full time on this."

SRP opened its Office of International Affairs in January 1984. Its mission included coordinating tours of SRP facilities for foreign visitors, arranging on-the-job training programs, and developing further employee exchange programs. In its first year, the office hosted more than four hundred visitors from fifty-three countries.

The outreach apparently was unique in the industry. In an employee Q&A in May 1985, Jack said, "I suspect we are the only utility in the United States that does have an Office of International Affairs."[2]

He had always believed in corporate responsibility and playing a role in community issues: Promoting the foreign policy objectives of the United States was a natural extension of his philosophy. The motives weren't without self-interest, as Jack explained: "I believe we have a continuing responsibility to support the Federal Government, because they've supported us."

PEEP, meanwhile, was a huge hit with the U.S. Agency for International Development, which named it the top exchange program of 1983. A second three-year version was launched in 1984.

Jack went back to Egypt in September 1985 for an executive management seminar for the Ministry of Irrigation. (The local English-language newspapers, obviously misled by the name of the Arizona utility, reported that the topic was how to use salt water for irrigation.[3]) Paul Ahler, assistant general manager of human resources, came along. After the plane landed in Cairo, Ahler recalled, there was a page on the intercom system for the SRP chief. The name Pfister was apparently too much of a challenge, so the call went out for, "Mr. Jack, Mr. Jack." It became a standing joke throughout the trip — although in a way, it symbolized Jack's accessibility and openness, traits that charmed his Egyptian hosts.[4]

The only glitch the SRP group hit in their presentations was humor, Ahler found. Whether it was culture or language barriers, the little jokes in their talks fell flat.

The visit included tours of Egyptian irrigation facilities, where the Americans immediately saw potential improvements. "One thing that was noticeable to all of us," Ahler said, "was that they didn't line their canals; they were losing tremendous amounts of water."

At the Aswan dam, the group was served lunch at the site manager's house, and when they signed the guest book, Ahler noticed that the previous visitor had been former President Richard Nixon.

Jack was a visionary, looking toward the future, Kirdar said. And he understood globalization early on: "We have to understand other cultures. A foreign country is not far away anymore."

Through the Office of International Affairs, students came from as far away as Sri Lanka to study the latest technology. SRP stood to gain, as well: One of Jack's guiding principles was that you learn by teaching.

An SRP brochure expressed how Jack calculated the benefits of international programs: "Experience, thus far, has shown that SRP participants return to the U.S. with a broader world perspective, better understanding of others' cultural values and exhibit a deeper appreciation for their country, jobs, and families. As the world improves, it becomes a better place for all, including SRP's shareholders, employees and customers."

Pumping Arizona Dry

In the West, they say, whiskey's for drinking and water's for fighting. There's no history of Arizona without tales of the battles over water. And there's no history of Arizona water without Jack Pfister's role as a deputy and peacemaker — sometimes well behind the scenes.

He shaped the 1980 Groundwater Management Act, a historic law that was a national milestone. And he played a vital role in getting it passed.

To do it, Jack had to persuade Salt River Project to give up power. He had to talk its agricultural interests, by far the largest users of groundwater in Arizona, into accepting a law that slapped new rules on them.

Despite its arid climate, Arizona sits on some immense aquifers. The catch is that the withdrawal rate far exceeds the amount replenished by sparse rainfall: Phoenix averages just eight inches of precipitation a year, compared with fifty in New York City and thirty-six in Chicago.

Yet groundwater pumping from the aquifers was historically a free-for-all. While the rights to surface water were based on past usage — summed up as "first in time, first in right" — landowners had unlimited rights to pump the water below their property. The law ignored the fact that neighbors were sitting on the same pool of water, and that one person's well could drain the one next door.

Proposals to control groundwater pumping stirred up fights, not action. With disproportionate clout at the Capitol, agricultural interests could fend off any limits on their access to water.

By mid-century, water levels were dropping at an alarming pace, accelerated by a string of dry years. In some areas, groundwater levels had plummeted more than 125 feet between 1940 and 1968. In 1948, Arizonans realized that the ground was sinking as the water was sucked out. Nearly 625 square miles around Eloy, a farm town about fifty miles south of Phoenix, had subsided by as much as 12.5 feet by 1977.

The earth began cracking open, too, as the water that had supported the soil was pumped away. The formation of fissures, which could be forty or more feet across and extend for miles, rose dramatically during the 1950s. A 425-foot gap opened up in a subdivision that was under construction in Paradise Valley in 1980.

What little oversight Arizona exercised over water was spotty and fragmented. One agency, eventually known as the Arizona Water Commission, mostly focused on battles over the Colorado River and other interstate issues. The State Land Department theoretically handled water rights. But when Bruce Babbitt took office as governor in 1978, he learned to his amazement that the state's de facto water department was Salt River Project. It was, he thought, "a really odd way to run a modern state government."[1] He knew it would have to change.

And so did Jack. As with flood control, he knew that SRP had two choices: participate and try to shape the rules or resist and have them imposed on the utility.

The spur to action was a court decision that threatened the water supplies of mines and cities. Farmers Investment Company (FICO), a pecan grower in southern Arizona, sued to prohibit mining companies from pumping near its orchards and then transporting the water to their operations several miles away. The city of Tucson, which was pumping in the same area, intervened, claiming the mines could pollute groundwater. The Arizona Supreme Court ruled in 1976 in favor of FICO and held that neither the mines nor Tucson could move water out of the area.

Agriculture won the battle, some said, but lost the war. The mines and cities now had a common goal: change Arizona's inadequate, outdated groundwater laws.

The Legislature started the painful process in 1977 by forming a groundwater commission that included lawmakers and representatives of three big water users: cities, mines, and agriculture. It would be unthinkable today, but no one from any environmental group was included.

To represent SRP, Jack chose one of his top executives, Leroy Michael. "It was a good choice," Babbitt said, "because Leroy was absolutely loyal to Jack and really had a great technical understanding of what was at issue." He was also a sharp, calm negotiator. An attorney from Jennings Strouss represented SRP's legal interests: Jon Kyl, who would go on to become a U.S. senator.

Jack's strategy was to position SRP as the overall representative of agriculture. If traditional farming interests became too involved, he worried, they would certainly derail the compromises that had to be made. "It was really a canny kind of insight," Babbitt said: Protecting the farmers by shutting them out.

Jack himself stayed away from the negotiating table to keep SRP's role low profile. Of course, he had no intention of ignoring the debate. He was on the phone every night with Kyl and Michael. "He wanted to know everything about it," Kyl said. "He mastered it as well as anybody. We ran all the major decisions by him."[2]

The basic questions were: Who gets water? How much? And under what conditions?

The commission held statewide hearings in 1979 and issued some overarching "concepts for agreement." But the process stalled out in November. So Babbitt resorted to theatrics. At his urging, Interior Secretary Cecil Andrus threatened to hold up the Central Arizona Project if Arizona didn't put some controls on groundwater use.

Over the next half year, a "rump group" of the major interests holed up in a conference room, with Babbitt presiding. Once again, Michael and Kyl were major players, reporting back to Jack, who then informed the SRP board. The rump group "had a lot of ups and downs," Jack remembered.[3] A big sticking point was how much control to impose on pumping. Agriculture wanted less, the cities wanted more.

Several times, the negotiations broke down over those city vs. farm issues. Then Jack would get together with Jack DeBolske, executive director of the League of Arizona Cities and Towns, his old pal from lobbying days back at the Capitol.

They met for breakfast at a place called Humpty Dumpty in central Phoenix. "We would sit down and talk about things I was having heartburn over," DeBolske recalled. Then they would hash out ways to get the talks on track again.[4]

Jack with House Majority Leader Burton Barr.

At long last, in June, a deal emerged. Three of Arizona's legendary legislative leaders had helped shepherd the process: Senate President Stan Turley, House Majority Leader Burton Barr, both Republicans, and Democratic Senate Minority Leader Alfredo Gutierrez. Now the Groundwater Management Act raced through a seven-hour special session of the Legislature on June 11, 1980. Babbitt signed it the next day.

The law created three levels of water management: general state-wide measures, more intense steps in what were dubbed Irrigation Non-expansion Areas, and the highest level of controls in the newly created category of Active Management Areas.

Three of the AMAs were urban areas — Phoenix, Prescott, and Tucson — where the goal by 2025 was to achieve "safe yield," pumping no more water out of the ground than nature could replenish. In the fourth AMA, the farming region of Pinal County, the goal was to preserve its agricultural economy for as long as possible, while also allowing for non-irrigation uses.

The requirements for the AMAs included comprehensive water plans, conservation targets, no new development without proof of a 100-year assured supply of water, and a ban on irrigation for new agricultural land.

The law created the Arizona Department of Water Resources to ride herd on water use and protect the state's share of the Colorado River. As director, Babbitt made the logical choice of Wesley Steiner, who headed the more limited Arizona Water Commission and was heavily involved in the negotiations.

The farmers were giving up a lot. Jack had to show the board what they were getting. His major selling point was the restriction on development without an assured water supply. SRP's big fear at the time was the growth outside its boundaries. If those homes and businesses ever ran out of water, they were certain to come after SRP's supply.

The law was "a victory for agriculture," in Jack's view. "I've always felt that it was a relatively balanced approach," he said, that didn't impose any greater burdens on agriculture than on cities.

It's one thing to pass a law and another to make it work. SRP board members accepted the Groundwater Management Act at first. But once they saw how it was rolling out, they resisted implementing it.

"I don't think any of us really fully understood what the total impact was going to be," Jack said.

Management, trying to follow the law, was increasingly at odds with the board. "They felt we were not aggressive enough in protecting the farms from unreasonable intrusions by the Department of Water Resources," he said. He had to fend off angry board members who wanted to get the law declared unconstitutional.

Some eight years later, Jack's old mentor, Rod McMullin, had become a loud critic, complaining passionately about SRP management's failure to protect farmers.

The legislation, as Jack and everyone else involved recognized, had plenty of flaws. Over the years, it was tuned up and expanded. It's still a work in progress, and the goal of "safe yield" is still far away.

Whatever its shortcomings, however, the Groundwater Management Act was a national landmark, the most comprehensive attempt that any state had taken to ensure its future water supplies. The Ford Foundation in 1986 named it as one of the ten most innovative programs in state and local government.

Max Sherman, a member of the National Committee on Innovations in State and Local Government, gave this assessment: "No other state has tried to manage its water resources so comprehensively. Arizona built a consensus around its policy and then followed through to make it work in practice."[5]

Looking back, Kyl said, "You can't say that without Jack it wouldn't have happened. But you can say that he was among the people that made it happen."

Taking care of the quantity of water in Arizona was just half the job, though. The state had to ensure the quality of water, too.

Jack had to lead another revolution in thinking at SRP. "Historically we had not thought we were in the water quality business," Jack said, "but in fact, we really were." Jack was indispensable in putting together another historic water law for Arizona, this one protecting quality.

Arizonans used to assume that water pumped out of the ground was pure and safe. But through much of the twentieth century, industrial and manufacturing operations dumped waste with little attention to whether poisons would seep into the aquifers below.

In 1981, the state discovered vast areas of contamination — not in remote spots, but in the middle of modern urban areas around wells that supplied drinking water. The noxious brew mixed into groundwater included toxic chemicals and suspected carcinogens, from chromium and chloroform to trichloroethylene (TCE) and benzene.

An enormous plume of tainted groundwater at the Motorola plant in Phoenix extended eight miles west, across the center of downtown. Toxins-laced water was detected underneath ten square miles around Tucson International Airport and thirty-five square miles centered on the western Phoenix suburb of Goodyear.

Dozens of wells had to be shut down that year, and the ugly discoveries of pollution just kept coming. Many of these places were soon declared federal Superfund sites.

Shocked Arizonans realized this most basic resource of life had virtually no protection. "As the pollutants were percolating down, political discontent was bubbling up," Jack said.

State law was murky and inadequate. Two institutions were in a tug-of-war for authority: the Department of Health Services and the Water Quality Control Council, whose thirteen members came from state agencies and major economic sectors.

Support was divided along political and ideological lines. Republican legislative leaders and the business community generally backed the water council. Gov. Bruce Babbitt, Democratic leaders, Tucson Republicans, and several environmental groups favored the health department.

When the Department of Health Services developed an industrial permitting program on its own, the Arizona Chamber of Commerce sued to stop it.

As the 1980s wore on, the Legislature repeatedly failed to pass a comprehensive water-quality bill. Finally, exasperated environmental and public-interest groups launched a drive to put an anti-pollution initiative on the November 1986 ballot. No one wanted to be on the wrong side of the dirty-water issue, and resolutions of support poured in. That focused the attention of the business community. So did the Chamber's loss of its legal challenge to the health department permit plan.

House Majority Leader Barr was getting ready to run for governor and didn't want an anti-pollution initiative, bound to be a focal point for critics, on the ballot with him. Tucson Republican Rep. Larry Hawke, who had been fruitlessly pushing water-quality legislation, brought together a variety of stakeholders to draft a bill. Health Services had its own proposal. And of course, there was the language of the initiative.

Now there was a bipartisan push to find a version that conservatives as well as conservationists would support. It had to be ready before the Legislature adjourned at the end of April.

In January, Babbitt created the Ad Hoc Water Quality Committee to do the job, and Jack was among its nineteen members. The group would hold weekly public meetings and, Babbitt pledged, do all its work in the open.

At its first meeting, Jack made a startling suggestion: Maybe Arizona should create its own state environmental protection agency. A single new department would then be responsible for clean air, water quality, and hazardous wastes.

The idea didn't get much notice as participants continued the old debate over the water commission, which some wanted to abolish, vs. the health department.

Babbitt divided the committee into two groups. He headed one, which tackled enforcement and process issues. Jack led the other, which dealt with the technical questions.

By late March, the public process still hadn't produced a bill. The participants complained that they were accused of selling out whenever they tried to compromise. So, just as with the water-quality legislation, the talks moved behind closed door to work out the difficult compromises.

Over the next month, both Babbitt and Jack used their talents for finding common ground to settle a long list of contentious points. Those included whether to require all underground water to meet drinking-water standards (yes) and whether farmers would need permits to use fertilizers and pesticides (no, but they would still be regulated).

Senate Minority Leader Alfredo Gutierrez and Jack.

They cleared the last hurdles on April 18. Potential polluters would have to get permits and regulate their discharges under a new aquifer-protection program. The notion that Jack had floated, a new state Department of Environmental Quality, was now a reality.

Senate minority leader Gutierrez, who was a major player in the negotiations, gave the credit for reaching a final deal to the leadership of two people: Babbitt and Jack.[6]

Roger Ferland, a lawyer representing Magma Copper Co. who helped draft the law, said, "Pfister was the genesis of the single thing that made this the strongest water-quality bill we've ever had."[7]

He was referring to Jack's proposal that companies be required to use the "best available technology" to limit waste discharges. That language, now familiar in anti-pollution law, ensures that improved techniques will continue to be added to groundwater protection.

But the bill also buffered the impact on agriculture and mines. A separate agricultural advisory committee would develop the best management practices for applying nitrate fertilizers and running feedlots. "Economic feasibility" would be taken into account. Older facilities, which basically meant mines, would have less regulation of discharges that affected the aesthetic quality of water, such as dissolved solids.

The Arizona Environmental Quality Act took effect on August 13, 1986. It was another landmark in safeguarding water. The *Los Angeles Times* called it "the nation's toughest law to protect underground water."

Jack's motto at SRP could have been the quotation attributed to Seneca: "The fates lead the willing and drag the unwilling."

"It's always been my thesis that if we didn't manage the water prudently, that we would lose the opportunity to do that," he said in his 1991 oral history. "If we weren't responsive to the changing standards of either flood control or water management or water quality, that someone would step in and do it for us. The Board, I think, has reluctantly, and sometimes grudgingly, accepted that."

The Tribes Are Thirsty

Most Arizonans didn't know it, but their water supply was at risk in the early 1980s. Indian tribes were pressing to end a century's failure to resolve their legal water rights. The future of the state and large parts of the rest of the West could be drastically changed by the outcome.

Arizonans were going to be awfully thirsty, Jack pointed out, if the tribes got all the water they claimed. The claims of just three of the state's twenty-one Indian reservations added up to more than half the state's predicted water consumption.[1]

Locally, tribal claims exceeded the capacity of the Salt and Verde rivers. "Those claims were grossly exaggerated," Jack said in his 1991 oral history, "but they represent the type of risks that water users in metropolitan Phoenix were exposed to."[2]

With its vast watershed and broad service area bordering or encompassing several reservations, Salt River Project was bound to get caught up in battles over tribal water rights. Even though he was a lawyer, Jack saw that the worst place to solve these water issues was in the courtroom. The adversarial nature of legal action was sure to create bad blood that would make future cooperation difficult. The process could stall over legal maneuvers and details that had no real relevance to finding a solution. Every decision was likely to be appealed. And the potential downside was heart-stopping: The worst-case scenario would be a disastrous reduction in water supplies for SRP and Arizona.

Now was the time for the negotiation, compromise, and search for common ground that were Jack's preferred approach to problems. He would personally end up at the table to work out a historic tribal water settlement.

◇◇◇

The U.S. Supreme Court was looking at a reservation in Montana, not Arizona, when it laid down the basic rule for tribal water rights in 1908. In what became known as the Winters Doctrine, the court held that when the federal government established reservations, it also clearly intended to reserve enough water for Indians to use the land.

The decision "seemed eminently logical and humane," Jack wrote in an article about tribal water rights.[3] But the Supreme Court didn't explain how to calculate the amount of water tribes should get.

Over the years, the federal government did nothing to settle Indian water rights. Instead, it added to the future conflict by encouraging non-Indians to settle, farm, and mine in the West. It built massive reclamation projects to provide low-cost water to non-Indians.

After World War II, tribes moved toward self-governance; there was more legal clarity about the reservations' status as quasi-sovereign nations. Now Native Americans were asserting their rights to water.

"Nowhere is this situation more significant than in the state of Arizona," Jack wrote. Nearly a third of the state's seventy-three million acres was reservation land but virtually all of the state's water was going to industry, non-Indian agriculture, cities, and towns. A showdown was coming.

SRP had reason to be concerned after Supreme Court rulings in the early 1960s held that the small tribes along the Colorado River were entitled to enough water for all the practicably irrigable acreage on their reservations.

When Arizona tribes applied that formula to their own reservations, the numbers were enormous. Mathematically, the Indian claims couldn't be resolved without compromise, Jack said. "If Arizona Indian communities were to receive all the water they claim…there would be no water for non-Indians and, ironically, not enough water to satisfy completely all the Indian claims."

SRP was pulled into the struggle over Indian water rights when the White Mountain Apache Tribe decided to build a dam on one of the tributaries to the Salt River in the early 1960s. The dam would capture some of the water that would otherwise flow into SRP reservoirs, and the utility sued to stop the tribe. But the lawsuit was dismissed over jurisdiction issues.

SRP wasn't going to be caught off guard again. It began extensive research on tribal rights, assembling an arsenal of legal arguments and experts. Jack wanted to work from a position of strength. "I've always believed that you needed to resolve these [issues] by negotiation rather than by litigation, but that the best way to make certain that you could do that was to be very well prepared for litigation," he explained in his 1991 oral history.

When the Groundwater Management Act passed in 1980, Jack saw one more reason to negotiate water issues with tribes. The Phoenix metro area — often called the Valley of the Sun, or just the Valley — was required to reach "safe yield" by 2025, pumping no more groundwater than Nature could replenish. But that would be impossible, even with the most stringent and painful conservation measures, if the tribes had free rein to increase their pumping.

As 1983 was winding down, SRP had already spent more than $1 million developing a historical and legal basis for defending the water rights of its shareholders. Now, some of the state's Colorado River allotment might be in jeopardy, too. The tribes were grumbling about the amount of water from the Central Arizona Project aqueduct that the Interior Secretary had reserved for them.

The most pressing claims were from the Salt River Pima-Maricopa Indian Community. And the most contentious point was its exclusion from the SRP system. The southern half of the reservation, which runs along the east side of Scottsdale, lies within SRP's service territory. Yet SRP had never recognized the rights of those tribal lands to a share of the water stored in SRP reservoirs.

Talks had repeatedly broken down over the issue. But Jack was determined to find a path that avoided litigation. On September 30, 1983, he sent a letter to Gerald Anton, the president of the Salt River Pima-Maricopa Indian Community, saying SRP wished to resume negotiations.

It could hardly have been encouraging when the tribe's next step was to sue SRP — although the dispute wasn't over water, but about alleged overcharges for electricity. In any case, Jack was undaunted. As a show of good faith, he offered to contact the federal Bureau of Reclamation to help resolve some issues the tribe had with delivery of water from the Central Arizona Project.

Richard Wilks represented the Salt River Pima-Maricopa community in the negotiations. Various SRP staff participated, he said, but Jack himself was nearly always at the table. "He stood hard for the rights of the Salt River Project," Wilks said, "but was willing to understand the position of the Salt River Indian Community."[4]

A couple of SRP staff members, however, were bullheaded corporate veterans, Wilks said, "who weren't going to settle if the world came to an end." Jack would have to put the brakes on them at times to keep from derailing the talks.

Jack was under a lot of pressure from the SRP board, dominated by farmers, to resist giving an inch. He had to show them that the tribe had the legal muscle to demand more than a few token drops of water. What he could do was bargain for the best deal he could get while still resolving the Indian community's claims.

At one point, early in the process, tribal President Anton and attorney Wilks went to Washington, D.C., where they had an appointment with Arizona's then-Congressman John McCain. When they arrived at his office, by coincidence a contingent from SRP, including Jack, was already waiting in the vestibule. After a while, McCain popped his head out the door and told Anton and Wilks, "Come on in."

"We were here first," the SRP group protested.

McCain looked back at the pair from the tribe and repeated, "Come on in."

One of the SRP crowd — it might have been Jack — remarked, "Somebody's telling us we're not the boss."

It was a hard lesson for Jack to drive home to an SRP board used to getting its way.

In 1984, a year after his original letter, Jack sent a second letter to tribal President Anton outlining a possible settlement. The points were mostly ones SRP had brought up before, but this time Jack added a twist: an offer to discuss how the tribe might receive SRP water. Anton responded that the delivery of SRP water should be the primary focus going forward.[5]

The back and forth had started. The plodding pace would have made the average executive climb the walls. But Jack, with seemingly infinite patience, was prepared to play the long game.

After another round of meetings, the Salt River community made its own proposal in June 1985. A key point was to give tribal land in SRP territory the same amount of water as other properties there.

SRP's counterproposal some six months later went even further. Reservation land wouldn't just obtain water, it would also be treated like any other land within SRP boundaries. That meant the Indian community would be able to store water behind SRP dams, a critical management tool.

Wilks knew Jack had plenty of other responsibilities at SRP, but the negotiations were so complex and demanding, "I couldn't conceive he was doing much else. His focus was very clear."

As if that weren't enough, around this time, SRP also began negotiating a water settlement with the Fort McDowell Yavapai Nation, north of the Salt River reservation.

But another water issue had been resolved, and it was time to celebrate. Arizona had struck a deal with the federal government to use local money to help pay for dam projects that were essential for flood control and storage for the Central Arizona Project. Otherwise the projects, known collectively as Plan 6, would have been delayed for decades while awaiting full federal funding.

Jack flew to Washington, joining a host of other Arizona political, water, and tribal leaders at the signing ceremony with Interior Secretary Donald Hodel on April 15, 1986. Michael Clinton, a veteran federal negotiator, was among the Interior Department officials at the event. At one point, he recalled, Salt River Pima-Maricopa President Anton broke into the festive mood with a complaint: His tribe's water rights still hadn't been resolved.

Jack responded, "Well, that's something we're going to put our attention to immediately." Later that day, Clinton was called into the Interior Secretary's office for a meeting with Hodel and Jack. He was being sent, at Jack's request, to help lead the negotiations, which were bogging down again.[6]

Clinton's presence revived the discussions and defused tensions. He had long experience in tribal water talks and knew how to spot the key points in a thicket of legal complications.

The old decision on calculating the water owed to the Colorado River tribes was based on inefficient flood irrigation. Clinton made a crucial breakthrough by suggesting that the Salt River deal should include funding for state-of-the-art irrigation techniques on the reservation — and the tribe's water needs should be calculated on that basis. "Jack bought into that idea wholeheartedly," Clinton said.

Once an amount was agreed on, the next hurdle was the source of the water. One of Jack's main concerns from the very start was the finite quantity of water available in Arizona.

Despite his sympathy for tribes and their water rights, Jack was deeply concerned about the prospect of taking water away from people who had used it in good faith. No plan would ever be acceptable to them if it fulfilled Indian water rights by draining the supply of non-tribal businesses, homes, and farms.

Jack had wanted to get cities in metro Phoenix involved in the negotiations from the start. Local city officials, he later grumbled, were quite happy to let SRP do the heavy lifting on water claims. However, that was Clinton's preference, too. He wanted the simplicity of having only the two most important players at the table.

But the cities had more flexible ways to handle water, and they could find ways to supply the tribes. SRP and Salt River Pima-Maricopa reached a broad conceptual agreement on the overall quantity of water the tribe would get. Phoenix and other cities had to be brought into the discussions, along with a water district in the southeast Valley, to hash out where the water would come from.

The ultimate solution included an intricate set of exchanges. Through various mechanisms, the tribe would get water from the Salt and Verde rivers, the Central Arizona Project, and wells — plus $3 million from Arizona and $47 million from the federal government for the tribal community's trust fund.

Groundwater would be protected. The tribe was required to achieve safe-yield when the adjoining part of the Phoenix area did so. The tribe got some income and eased the pressure on water supplies by agreeing to lease its Central Arizona Project allocation to local cities until 2099.

Within a couple of years, there was a settlement with Fort McDowell Indian community, as well. The deal included water from the Verde River and the Central Arizona Project, which the tribe was allowed to lease, and a guaranteed minimum flow on the Lower Verde River, which runs through the reservation. A development fund was created with $23 million from the U.S. and $2 million from the state.

Congress approved the Salt River-Pima Maricopa Indian Community Water Rights Settlement Act in 1988 and the Fort McDowell agreement in 1990.

For Jack, the deals underlined his conviction that negotiation was more effective than litigation. Arlinda Locklear, who represented the Fort McDowell community in the discussions, said, "I don't think anybody in this negotiation feels as if they got burned. And that is unusual, particularly in something this complex."[7]

The Interior Department pulled Clinton back before the settlement was complete. He retired then and was hired by Arizona interests to lobby Congress for the appropriations to carry out the settlements. Jack was at his shoulder in many of the Washington meetings. "He was a champion," Clinton said, persuading Congress that the federal government couldn't duck its financial responsibilities. The money was appropriated.

From his years of negotiations, Clinton knew how to tell the blowhards from the leaders, and the obstructionists from the builders. The Salt River Pima-Maricopa agreement removed a cloud of uncertainty off Western business, farming, industry, and cities.

In Clinton's view, "The settlement would not have happened without Jack's leadership."

The Mathematics of Tuition

Jack Pfister as Marie Antoinette? When Jack was appointed to the governing board for Arizona universities, he had no idea how turbulent the next eight years would be. Or how editorial pages would lampoon him as the imperious French queen. Or how thoroughly he would enjoy the chance to shape university education in Arizona.

Unlike many other states, Arizona has just three public institutions of higher learning: Arizona State University with its main campus in Tempe, Northern Arizona University in Flagstaff, and the University of Arizona in Tucson. The Arizona Board of Regents sets policy for them in everything from finances to academics to student affairs. In the 1980s the board had eight voting members, who served staggered eight-year terms, plus a nonvoting student regent and the governor and the state superintendent of public instruction as ex officio members.

Jack Pfister served on the Arizona Board of Regents from 1982 to 1990.

Jack had a passion for education, and he'd developed a personal relationship with Gov. Bruce Babbitt. Wonks at heart, they talked about the details of state government with the enthusiasm of sports fans poring over scores. So when two members wrapped up terms on the Arizona Board of Regents in 1982, Jack was a natural choice to fill one of the vacant seats. Not that he even hinted to Babbitt that he wanted the post.

"No, that would have been totally outside his persona to ask for something personally," Babbitt said. "It just wasn't part of him. It was a merit selection, it really was, backlit by my personal admiration for the guy."[1]

The governor picked a Democrat, Tucson medical administrator Don Shropshire, as a counterweight to Jack's Republican record. (When reporters asked, however, it turned out that both had donated to Babbitt's election campaign in 1978.[2]) The Senate confirmed them without a hitch.

Tuition was a battle throughout Jack's term. The Arizona Constitution requires that "instruction furnished shall be as nearly free as possible" at universities. But legislators were starting to resent how much the state was spending on higher education.

Historically, the regents had no particular method for setting tuition. They picked numbers out of the air based on political pressure. They went years without raising tuition in the 1970s, despite rising inflation.

Then, with the Legislature threatening in 1980 to take over the authority to set tuition, the regents adopted the "fair share formula." Arizona residents would pay 20 percent of the average cost of their education, and the state would pick up the rest. Non-residents would pay 85 percent at Arizona State University and University of Arizona and 75 percent at Northern Arizona University.

Soon after Jack joined the board of regents, the formula began falling apart. The state budget was facing a $200 million hole in 1983, and legislators expected the regents to help fill it. They wanted the board to raise registration fees and increase out-of-state tuition.

Jack sized up the situation diplomatically, but with a note of caution to his fellow regents. "The Legislature is faced with the toughest job they've had in recent history," he said. "I think they're trying to say to the regents they need our help. If we turn our backs on them now, they're going to remember that."[3]

As chair of the regents' finance subcommittee, Jack knew the question wasn't "if" but "how much?" The amount was tricky, with a risk of being counterproductive. If the cost of attending the universities went too high, enrollment would drop — and total revenues would be lower, not higher. Arizona had hiked tuition for out-of-state students by more than 50 percent in the past three years, and enrollment had dropped 16 percent.

On the other hand, if the regents made only a token increase, the Legislature could decide to set tuition rates itself. With that sword hanging over their heads, the regents approved a $95 one-time surcharge on all students in April 1983. That would boost annual in-state tuition to $850.[4]

Cost wasn't the only issue. The overall quality of Arizona's universities concerned Jack. He'd seen all three up close, having graduated from UofA, teaching as an adjunct professor at ASU, and sending his son, Scott, to NAU. "The universities were not very demanding of their students," he said. "There was a pre-occupation with numbers as opposed to quality."[5]

The regents were trying to set the stage for Arizona to become a high-tech mecca. They wanted freshmen to walk onto campus with sharper skills, which meant more stringent high school course requirements for admission to state universities.

The board met in May 1983 to decide on the standards.[6] Jack lobbied against requiring geometry, saying, "I am not as certain it will be relevant in the future. I don't think we ought to be so restrictive."

Representatives from the UofA and ASU strongly defended geometry as "the foundation of thinking" and "useful no matter what you do."

The arguments were persuasive, and one of his fellow regents wondered if Jack wanted to refute them. In an elegant twist that mathematicians could appreciate, Jack answered simply, "Q.E.D." — the abbreviation for quod erat demonstrandum, "which was to be demonstrated," the Latin phrase traditionally put at the end of a mathematical proof.

While conceding on geometry, he headed off a proposal to drop the fourth year of English. Jack argued that a fourth year reinforced composition and writing skills that were critically needed in the real world. "As one who reads lot of reports, this is one of the major deficiencies," he observed.

The board unanimously approved the new requirements for incoming freshmen: four years of English, three years of math, two years of laboratory science, and two years of social studies.

ASU wanted to open a campus on the west side of the Phoenix metro area. In a growing state, the need was there, and at the November 1983 regents' meeting, Jack argued, "I feel it's appropriate to get started."[7] He pointed out ways around the most significant obstacles. The short-term negative impact on the other two universities could be offset through tuition policies, curriculum, and facilities that enhanced their competitive position. At the Legislature, funding for the west-side campus could be a separate proposal to keep it from impacting the universities' overall budget request.

ASU West got legislative approval in 1984 and in the next three decades grew to serve 9,000 students.

Meanwhile, there was a tug of war over money. While they couldn't avoid renewing the one-year surcharge for the 1984-85 academic year, the regents balked at increasing tuition. They put off making a decision in July 1983, considered holding the line in November, and rejected rate hikes proposed by the university presidents in December.

The debate spilled over into 1984. Governor Babbitt, facing another year of budget deficits, wanted to set tuition at $950. But he included a note of caution: "If we raise tuition, and legislators use the additional revenue to fund prisons and health care, then I think that regents would feel that they have not fulfilled their obligation to students, parents, and the universities."[8]

Jack was working his political connections, chatting with key legislators, and fighting a bill that would mandate tuition levels for the next three years, stripping the board of its power.

The regents finally accepted the inevitable in late May 1984 and went with an in-state tuition of $950 and a non-resident cost of $3,700 for UofA and ASU. Out-of-state tuition was now higher than that of 70 percent of comparable universities. "We're very rapidly reaching the point where we're going to price ourselves out of the market," Jack warned.[9]

A week later, Jack succeeded in fending off the Legislature's tuition power grab. Defending the regents' track record in holding down tuition, he told the House Education Committee that with his modest background, "Were it not for the bargain of an education in the state of Arizona, I would not be where I am today."

University funding was a hard sell to many legislators, however. Rep. Jim Cooper, a Mesa Republican and a dairy farmer, had a counterpoint to Jack's story: He himself had chosen to drop out of school and go into business. "I'm better off financially," he said, "and I'm chairman of the education committee. Sometimes it pays not to go to school."[10]

When it came time to set tuition for 1985-86, the regents were closing in on the psychologically significant $1,000 mark for in-state students. They held off. The Arizona Students Association took credit for persuading them to keep the increase to $40.

But in September 1985, the board learned that Arizona's in-state tuition of $990 was the eighth-lowest among the major "flagship" universities in the fifty states. The university presidents recommended the largest jump ever: $146. Jack was on board. He favored offsetting the extra burden on low-income students by increasing financial aid.

"I believe, by any standard, our tuition is too low," he said. Between 1972 and 1978, tuition had gone up just once, by 9 percent, while inflation was running 45 percent. "We've never caught up," he said.[11]

The regents were committed to keeping tuition levels in the bottom third of U.S. colleges and universities. So they still had plenty of room for an increase. And they hoped that by having students carry more of the burden, the Legislature would be more willing to put more money into the universities.

As a regent, Jack had to balance two of his core beliefs about higher education: It should be accessible to all, and it should be top notch.

More than one hundred students and parents jammed the November 7 regents meeting in Tucson to protest the size of the tuition hike. The next day, the regents voted yes. Arguing for the increase, Jack said, "I speak on behalf of the students who are going to have to suffer from an inferior education. I believe the time has come for those students to be represented as well."[12]

In-state students would pay $1,136. Nonresidents would be charged $4,261 at ASU and UofA.

A *Phoenix Gazette* editorial compared him with an arrogant Marie Antoinette. The *State Press*, ASU's student-produced newspaper, ran a cartoon showing a whistling figure in a suit and tie, walking down the street with a briefcase labeled "students," cheerfully unaware of the masked hoodlum — labeled Pfister — waiting to mug him around the next corner.[13]

The cartoon ignored the cushion for those who might have been priced out of school. The tuition hike was expected to bring in an extra $12.3 million, and $2.5 million of that would expand the tuition-waiver program for low-income students.

Jack was always the practical voice on the board, weighing how their actions would play out. He was careful to play within the rules, although he was also a master at using them to strategic advantage. Those traits prompted him to move cautiously even on — and perhaps particularly on — issues that resonated with his philosophy and ethical standards.

One was the fight to end apartheid, South Africa's repugnant and brutal system of keeping races separate.

When Jack joined the Board of Regents, a global move to put economic pressure on the South African government was gaining momentum. A major tool was "disinvestment," getting companies and institutions to pull their money and operations out of South Africa. In mid-1985, groups at ASU and UofA called on the universities to cut all their ties to South Africa by the end of the year.

More than thirty-five other colleges and universities had already acted, adopting policies denouncing apartheid and shrinking or ending their investments with South African ties.

Arizona's three universities had endowments of $31 million, with just over 10 percent invested in companies doing business in South Africa. The yardstick for investing the money had always been financial: balancing yield and safety. At the end of August, the regents' finance committee spent three hours debating whether to consider moral issues, as well.

More than twenty protesters picketed the meeting, with their spokesman arguing that "Public monies should not be used to prop up institutionalized slavery and discrimination."

But Jack worried about setting a precedent if the regents responded to a single issue. "I'm personally opposed to a policy that deals only with South Africa," he said. "It would be an invitation to pressure groups" on a variety of other issues.[14]

The committee agreed and decided to recommend adding moral and ethical concerns to the universities' investment policy, without targeting South Africa.

The question came to the full board in September. And this time Jack lost. With one member absent, the board voted 4-3 to order universities to cut their financial ties as soon as possible to U.S. companies doing business in South Africa.

Board President Donald Pitt cast the deciding vote, saying, "I think the U.S. has to do everything it can to support the elimination of apartheid. We should put freedom above dollars."[15]

Saving the Affordable Option

Maricopa County has the largest system of community colleges in America. They're supposed to be the affordable option for higher education, offering technical instruction, workforce training, and a two-year university transfer program. But in the 1980s, the costs were going up and up. Over the decade, tuition skyrocketed more than 600 percent, while scholarships were scarce.

Students were getting priced out of an education, and that was just unacceptable to Jack. He wasn't going to let finances close the door on willing students.

Jack had never forgotten how his father's insurance policy made higher education possible for Tad and him. "If it hadn't have been for that, I probably would not have gone to college," he said, "so I know how important a relatively small amount of money can be at a critical time."[1]

There was a vehicle to raise money: the Maricopa Community Colleges Foundation. It was low profile, though, with few assets. Universities and four-year colleges across the United States had a long track record of bringing in money through foundations, sometimes amassing huge endowments from alumni. But the concept was new for community colleges, which didn't cultivate the same relationship with former students.

Jack already had a connection with Maricopa Community Colleges — he'd served on the steering committee during its $75 million capital bond program in 1984. Now he stepped up to lead the system's first public fundraising drive.

The campaign needed a treasurer. So he turned to Dennis Mitchem, an accountant and fellow member of the Phoenix 40. Mitchem was dubious about the whole idea: "I said, 'Community colleges don't have foundations, and they don't do fundraising.'"[2]

Mitchem was already chairing two high-profile campaigns, United Way and an ultimately unsuccessful river development project called Rio Salado. "I didn't need another thing to do," he said. But Jack insisted: The public had to feel that the tax side of the fund drive was properly handled, and Mitchem's participation would be the seal of assurance.

"He appealed to my special skills and my interest," Mitchem said. "Jack was one of those people you really don't say no to. I bet you he got very few turndowns. He did his homework, he knew what he was doing, and he was right."

At the time Paul Elsner was chancellor of the community colleges. He had little experience in fundraising. "Jack often set up meetings with many major, potential donors, many of whom I would not have easily approached," he recalled.[3]

One challenge was to demonstrate that even a relatively small tuition could stop some people from going to college without a scholarship. So they made a video of a young mother who had a low-wage job and lived with her children in a car after losing their home. She managed to win one of the few scholarships to community college and eventually got a job as an accountant. Mitchem pitched several major corporations to make a contribution, and "I can remember that video being very, very effective."

The campaign ran from 1987 to 1989, with a public push at the end. The goal was to raise $4.5 million. It brought in $6 million.

Jack went on to join the foundation board, where he was an active, hands-on member and served a term as president, said Mitchem, who was also on the board.

Jack led the foundation's second fundraising drive, "Investing in Arizona," from 1997 to 1999. This time, the goal was $12 million, and once again the target was surpassed. The ten-campus Maricopa Community College District now had nearly 190,000 students in for-credit classes. The foundation's new injection of money would pay for up to 16,000 scholarships.

Jack and Pat made a personal commitment to the community colleges, as well. The Pfister Family Scholarship offers $1,000 to six students pursuing degrees in civil service. When Jack passed away in 2009, contributions were collected to support this fund. Five years later, it was endowed with more than $160,000.

Rough Riding for the Regents

Overall, everything was pretty quiet at the Board of Regents when Jack Pfister started a one-year term as president in May 1986. It was, as the movie cliché goes, too quiet. Outside the relative tranquility of the universities, Arizona's political leadership was in turmoil. State elections were approaching, and Gov. Bruce Babbitt, with his eye on the White House, had decided not to run for another term.

Democrats couldn't settle on a successor. They finally picked the superintendent of public schools to run for governor — then her defeated opponent in the primary election jumped back into the race as an independent in the general election. On the GOP side, the powerful House majority leader, Burton Barr, seemed to have a lock on the nomination. But Republicans unexpectedly nominated Evan Mecham, an auto dealer with ultra-conservative views who had made a second career out of running for office. With the Democratic vote split, Mecham won the election with 40 percent of the vote.

Days after he was sworn into office in January 1987, Mecham put Arizona in the national spotlight by abolishing the state holiday honoring Martin Luther King Jr. — just as he'd pledged to do. Now he was on a mission to fulfill his campaign promise to cut taxes and shrink government spending. And he saw the universities as fertile places to save money.

Suddenly Jack and the board were dealing with a governor who didn't take the value of higher education for granted. Mecham had attended Arizona State College (now Arizona State University) but dropped out to start his first car dealership. He wanted universities to de-emphasize research and concentrate on classroom instruction. He filled a vacant seat on the Board of Regents with someone he could count on for support: his main political fund-raiser, Ralph Watkins Jr.

With Babbitt, Jack had a governor who shared much of his philosophy and valued his opinions. The state's new chief executive was from another philosophical planet. While Jack spoke cautiously and diplomatically, Mecham was a verbal flame thrower. The new governor was prone to conspiracy theories and accused the state's "educationists," especially university officials, of using misinformation and scare tactics to create panic over his proposed funding for education.

Mecham saw himself as a crusader against overspending, and he castigated the regents. "The university presidents run the Board of Regents, instead of the other way around," he charged in February 1987. He was headed to a board meeting the next day and predicted he'd be as welcome there "as a polecat is at a slumber party."

Jack was far too savvy to take the governor's bait — he told a reporter he "respectfully disagrees" with Mecham — and he certainly wasn't going to treat anyone like a polecat. Jack's strategy wasn't to fight, but to bring people into the fold. So he made an oblique appeal to Mecham for teamwork. "A successful university system requires a collaborative effort between the board of regents and the presidents, rather than a confrontational effort," he said.[1]

Jack respectfully disagreed with Gov. Evan Mecham while serving on the Board of Regents. Sitting between the two is Congressman Bob Stump.

<>><>

But with Mecham, Jack couldn't avoid a confrontation.

The February meeting had been chugging along in Tucson with Mecham's usual provocative comments. The governor insisted that his proposal to cut university appropriations by 10 percent wasn't a reduction but "a steady state budget." He complained that too much money was going into research and said he was investigating how much time professors actually spent teaching.

Then Jack made the usual call to the audience for comments. Helen Cullen, a Massachusetts professor doing work at the University of Arizona, rose to be recognized. She looked mature and well-dressed. Jack decided to let her speak. Declining to come forward for a microphone, Cullen let loose.[2]

"Governor Mecham, you're coming through as an ass," she said. "Your handling of the Martin Luther King birthday was atrocious. You're cutting the budget of the universities. What are the students supposed to do? Go to Harvard?"

Mecham didn't respond. But his appointee, Watkins, jumped in, "Mr. Chairman, point of order."

"You look like a dried-up, old Bible-thumping preacher!" Cullen went on.

"Point of order, Mr. Chairman," Watkins called again.

"The students need help," Cullen continued. "The future is with public education. You act like an ass, but then you're a Republican."

"Helen," Jack broke in as Watkins spluttered another "Point of order!" But she was finished. "Thank you," she said, and sat down.

The whole incident lasted just a minute or two, and the meeting moved on.

But Mecham's allies were furious that Jack hadn't cut off Cullen immediately. State Rep. Bob Denny, a Litchfield Park Republican, shot off an outraged letter and passed it on to reporters. He wrote that "it was inexcusable for you — a community leader, a businessman, and president of the Board of Regents — to allow someone to attack anyone, let alone our governor, with that type of language in front of the Board of Regents."[3]

Apologize or resign, Denny demanded.

Jack wasn't going to do either one. Instead, he used the flap to explain his philosophy about public life: let people express their views, expect criticism, avoid fruitless confrontations, and develop a thick skin for nasty language.

The lesson of the 1960s, to Jack, was that it was better to let students talk directly to the regents than take to the streets in protest. The regents' policy was to let people give their opinions during the call to the audience. They had put up with nastier language than "ass," including being called racists and bigots.

In a letter to Denny, Jack wrote that seventeen years in the utility business had given him experience dealing with "rabid environmentalists, irate customers, angry flood victims, fired employees, and others who had hostile and passionate feelings they wanted to express." The best strategy, he had concluded, "is to permit these individuals to complete their statement and let their passions subside."

"Governor Mecham has indicated that he intends to be an active member of the Board of Regents," Jack wrote to Denny. "I welcome his participation in the work of the Board. However, active involvement inevitably includes exposure to offensive and abusive verbal statements. I cannot protect myself and others from the negative side of public participation. It goes with the territory."[4]

Mecham announced in January that he planned to seek repeal of a one-cent state sales tax adopted in 1983. That would take $250 million out of the state coffers — money needed for crucial programs, especially the universities. He claimed that repealing the sales tax would be a signal of the state's pro-business climate.

Jack wrote a memo to his fellow members of the Phoenix 40 urging them to launch a counterattack: "The governor and the Legislature need to know that a favorable business climate also includes an adequately-funded education system, including a university system that can respond to the needs of the state's business community."

A financial assault on higher education would be a "tragic error," Jack said, forcing faculty and staff reductions of five hundred at UofA and three hundred at ASU. The state should take a lesson from Texas, he said, where stagnant funding led to a "brain drain" of talented faculty.[5]

Business leaders across the state heard Jack's warning that Arizona would lose a competitive edge if it weakened higher education. Four major business organizations in Phoenix, Tucson, Mesa, and Flagstaff came out against repealing the tax if it would hurt higher education — the first time ever that they'd tackled a problem together, said representatives of the groups.[6] Legislators were persuaded to let the tax stand.

Mecham, for his part, had declared war on waste in government. His model was the Grace Commission, President Ronald Reagan's federal cost-control project. The governor's own mini-Grace Commission was going to pore over every aspect of state spending.

The last thing Jack could have wanted was for an ax-wielding governor to make recommendations about how the universities spend their dollars. So at the regents' March 1987 meeting, he proposed a task force to study the entire university system. He called for examining the balance between research and classroom instruction and ensuring "that we are getting the highest rate of return on our investment in education" — phrases sure to strike a chord with Mecham.

The regents unanimously approved the study, which would take fifteen months. It would be groundbreaking. While other states had created similar task forces, said Molly Broad, the board's executive director, Arizona had never before taken a comprehensive look at the quality of higher education.

At the meeting, Mecham was enthusiastically on board. "I'm delighted. Pleased. Ecstatic," he said.[7] But the governor quickly had second thoughts — especially after an unidentified university official boasted to the *Chronicle of Higher Education* that, "We outflanked him."

Now Mecham wanted to wrap the universities into his overall examination of efficiency in state government. He complained that the regents couldn't be objective when they were protecting their turf.

"We're not wanting a fight," Jack said, as the clash went public. But he held firm. As always, he bolstered his position with facts: Pulling the minutes of the meeting, he noted that Mecham had not only agreed with a separate university task force, but he also seconded the motion to create it.[8]

Jack decided to lead the Task Force on Excellence, Efficiency and Competitiveness himself.

He defended the panel against critics — notably Mecham and the editorial page of the *Phoenix Gazette* — who charged that it was stacked with university sympathizers. (The one obvious exception being the regent Mecham had named to the board, Ralph Watkins.)

But Jack took note of the newspaper's complaint of an "elitist bias" in a group with all Anglos and just one woman. "We stand corrected," he wrote in a letter to the editor in August, and within weeks he added a black Tucson businesswoman and the president of a Hispanic lawyers' association.

Jack had made it clear that he didn't expect to find a lot of waste. While the three universities had some inefficiencies, he said in a June speech to the Glendale Rotary Club, "They don't appear to be significant."

The regents had already directed the universities to review their curriculum, make courses more demanding, and focus on communication skills in every discipline. Entrance requirements were tougher, which would save money on remedial education.

Jack pitched universities as engines of economic development that would also provide intellectual and cultural leadership. Their researchers could find solutions to Arizona's problems, like air and water quality.[9]

The task force was a major production. Eight faculty members, drawn from the three universities, devoted themselves to it full time. The regents hired the former president of Oregon State University to serve as executive director and the management-consulting firm of Coopers & Lybrand to provide outside evaluations of the university system.

The final product, delivered in 1988, was mammoth, with fifty working papers and reports. The overall tab was $1.3 million, more than double the originally planned cost.

The result had nothing to do with the checklist of spending cuts that Mecham wanted. Under Jack's guidance, the focus had become how to strengthen the universities. Coopers & Lybrand had found no "buckets of fat" in the system, he said. In some areas, in fact, extra funding was needed to improve efficiency.

"Those who expected the report to give easy answers will be disappointed," Jack wrote in a September 25 guest column for the *Arizona Republic*.

There was one unexpected twist: some heavy criticism of the Board of Regents itself. The consultants found a string of organizational shortcoming, from hazy lines of responsibility to overlapping staff duties to a preoccupation with trivial issues. Skateboards were a case in point. At two different meetings, the regents discussed whether to ban skateboards at Arizona State University. (They did.) Meanwhile, they left lower-level administrators to deal with critical issues, such as developing information systems.

The task force made two dozen recommendations, and the top issue wasn't money, but minorities. While minorities made up 27 percent of Arizona's population in the college-going 18-24 age bracket, they were just 13 percent of the state's university students. This underrepresentation was a longstanding problem not just in Arizona, but around the nation.

The task force called for partnerships with local high schools, increased minority enrollment and graduation rates, and stepped-up recruitment of minority faculty and staff.

The recommendations in other areas — undergraduate education, access for rural students, and administration — were painted with equally broad-brush strokes. The ideas included creating a third campus for Arizona State University in central Phoenix and ratcheting up in-state admission standards at ASU and the University of Arizona.

Still, one of the legislators on the task force questioned the results. "Our initial goal was to find out if we got the best bang for our buck and I'm not sure we got that," said Republican Rep. Joe Lane. Mecham's mini-Grace Commission, on the other hand, had a $1.5 million budget and came up with possible savings of $250 million a year in other segments of state government.[10]

Touchdown Season

Jack wasn't much of a sports fan. He spent seven years in Tucson as an undergraduate and law student at the University of Arizona and yet, he admitted, "I have never been to a UofA football game."[1]

As a regent, he wanted to make sure that athletics didn't overwhelm academics. But when Arizona was trying to lure the Cardinals professional football team, Jack was a fan and helped coach the deal.

The state had been trying to snag a pro football team since 1974, when the National Football League passed over Phoenix and instead expanded to Tampa and Seattle.

When Jack became a regent in 1982, one of the first issues he faced was whether to let the Arizona Wranglers lease Sun Devil Stadium at Arizona State University. The Wranglers were part of the new United States Football League, which was gearing up to create a spring season of pro football after the NFL games were over.

Jack worked on the final negotiations, which led to a deal in October 1982. But the new league sputtered. The Wranglers became the Outlaws — and then disappeared.

In the meantime, Valley sports boosters had been courting owners who wanted to move their NFL teams. The lure was a domed stadium they hoped to build in downtown Phoenix. They had a serious flirtation with the Eagles, but Philadelphia managed to hang onto its team with a package of incentives that included building luxury boxes at the stadium. The importance of those skyboxes wasn't lost on Valley leaders.

The owner of the St. Louis Cardinals, Bill Bidwill, gave the Valley a look in 1985. Two years later, with St. Louis refusing to build him a larger stadium, he was seriously shopping.

Now the Valley had more bargaining power. Sun Devil Stadium had shown it could host a pro football team. It could be temporary home to the Cardinals while a domed stadium was under construction. If the regents gave their blessing.

ASU did the direct negotiations with Bidwill, with Jack acting as the board's liaison on the issue. His main concern was to protect the university's finances, making sure that not one dollar was diverted from education. By December, the package was about wrapped up. But skyboxes were a major sticking point. To finance the construction, they had to sell out. The Phoenix 40 held a retreat in Sedona to come up with a solution.

Jack rode herd on the discussion. "He was the moderator of all moderators," said Herman Chanen, founder of a construction company and a regent at the time. "He had the great ability of bringing people together, whether they agreed with each other or not, and keeping everybody calm."[2]

Building the sixty skyboxes would cost $6 million, and the leases would run $60,000 a year. The quick answer was for the executives at the retreat to invest in the boxes. But they had to be sold on the dual value of doing so: the overall payoff of bringing an NFL team to the Valley and the economic advantages of using the games to woo clients and reward employees.

Jack, as always, kept the conversation orderly and made sure everyone had a chance to speak. He had a knack for summing up a person's arguments. Then he'd call on others to give their reaction, said Bill Hogan, former general manager of Phoenix Newspapers Inc., who attended the retreat.[3]

At one point, Jack stepped out of the role of moderator and spoke as president of the Board of Regents. He laid out the pros and cons of the impact on ASU's football team.

In the end, a yellow pad was passed around so companies could sign up for skyboxes. The supply sold out then and there, Hogan recalled.

With the skybox deal, "Valley officials have tied the bow on their package for NFL Cardinals owner Bill Bidwill," trumpeted a December 17, 1987, article in the *Phoenix Gazette*.[4]

On January 15, 1988, Bidwill announced that the Cardinals were moving to Phoenix. The next day, the regents unanimously approved the terms of the stadium agreement. The exact wording of the lease still had to be worked out. Plus, Bidwill still needed the NFL's approval. But those were loose ends. The deal was done.

Then a dispute over booze threatened to blow it all up. The corporations and groups planning to lease the skyboxes expected to serve drinks during the games. But the regents prohibited the sale of alcohol at sports events, in line with the policy of the National Collegiate Athletic Association.

The solution seemed simple. In April, the Regents announced a proposal to exempt skyboxes from the alcohol ban during Cardinals games. But students and football fans were furious. They called the plan unfair, elitist, a double standard.

Arizona Republic cartoonist Steve Benson portrayed Jack as a wickedly funny Marie Antoinette, in all her ruffled and feathered splendor. Champagne class in hand, she pauses on a staircase to "the booze boxes," looks at the throngs protesting outside and sniffs, "Let them drink Coke." Jack was so tickled he got a copy from Benson, who signed it "to Jack — 'snob of the common man.'"

It was actually the second time the regents had considered a liquor loophole. The previous fall, UofA had asked to allow alcohol in the skyboxes planned for its own stadium — and then withdrew the request after a storm of protest.

This time, Jack warned, alcohol was a deal breaker. "The question is not just whether you serve alcohol but whether the Cardinals are going to be in Sun Devil Stadium," he said. The exemption was vital to leasing skyboxes, a critical chunk of the Cardinals' income stream.[5]

When the regents met in June, Jack focused on safety and security. "It's not an egalitarian thing in my mind," he said, "it's a control issue." Keeping tabs on skyboxes with a capacity of eighteen people each was possible, he said, while controlling a crowd of 70,000 was not.[6]

The regents voted 6-2 to allow the skybox exemption. A *Phoenix Gazette* editorial applauded the decision. But the *Arizona Republic* castigated Jack, "usually clear-headed," for catering to fat cats in plush, air-conditioned skyboxes.

With the liquor question settled, the regents approved a 30-year stadium lease with the Cardinals in June 1988, in the nick of time for the team to get ready for kickoff that fall.

As he accumulated the scars of budget and tuition fights, Jack realized that the universities needed other revenue streams. He became the regents' prime proponent of harnessing research to haul in dollars.

Businesses were less than thrilled, however, when competition came from the campus. UofA, which operated the state's medical school, triggered complaints because it was fitting and selling hearing aids. The university defused the issue by agreeing to buy hearing aids from private vendors.

But the rancor lingered. The Private Enterprise Review Board, which looked into accusations of unfair competition from state agencies, met with the regents in November 1988 to grouse about their slow response to complaints. Jack, confident as always in the calming power of communication, offered to set up annual meetings to go over problems. But he wasn't about to pull the universities out of the arena.

"We are in fact going to become more entrepreneurial," he said. "We need to accelerate turning scientific knowledge into practical products."[7]

Jack foresaw the economic reality. In 2012, UofA created Tech Launch Arizona to put its technology transfer and commercialization efforts into one organization. In fiscal 2014, the university chalked up 167 patent applications in fields ranging from optics to medicine and signed thirty-nine exclusive licenses to turn research into products

⟨⟩⟨⟩⟨⟩

The job description for university president could include "lightning rod." ASU President Russell Nelson had certainly drawn plenty of bolts since arriving in 1981. The business, journalism, and engineering programs all faced accreditation problems during his tenure.

One of his early achievements was restoring credibility to a scandal-plagued athletic department, which had been sanctioned for recruiting violations, phony grades, and doctored transcripts. But Nelson himself was criticized for overriding a faculty athletic representative's decision to bar a football star from playing during the fall 1987 season.

His toughest issue was minority access and recruitment. Arizona was in the first wave of states that would become "minority majority" in the next few decades, where more than half the population would be non-white. But a report in November 1987 put Arizona dead last among five Southwestern states in minority university enrollment. While nearly one in three Arizonans 15-19 years old came from a minority group, just 11 percent of the university student body did.

Hispanic leaders were clamoring for ASU, in particular, to do more — there wasn't a single Hispanic in a decision-making post. The School of Social Work was roiling with charges of racial discrimination.

Nelson pledged to make changes. In June 1988, he announced a twenty-one-point plan to bolster minority scholarships and faculty appointments. But it was too late. By coincidence the annual Arizona Chicano Conference was meeting at the same time. The prominent Hispanic business and political leaders there were fed up with ASU. They had one word for Nelson: Resign.

Jack was aghast. The regents and universities had spent as much time and energy on minority recruitment and retention as on any other subject, he said in a long letter published in the *Arizona Republic*.

"It is appropriate to express disappointment with progress and encourage improved performance," he wrote, "but to demand the resignation of one of our state's most effective and articulate spokesmen for affirmative action is counterproductive."[8]

As ASU headed into a new academic year in August 1988, Nelson told a stunned audience of faculty members that he was stepping down to live "a simpler, more private life." He would leave in June, giving the Board of Regents time to find his replacement.[9]

For Jack, it was a double loss: a highly successful university president and a good friend. Nelson oversaw a period of explosive growth. When he left, ASU was the nation's fifth-largest university. Nelson's accomplishments were right in line with Jack's vision for higher education in Arizona. He pushed academics over athletics, ignoring the howls of alumni boosters.

He led ASU far beyond its reputation as a party school. He boosted research funding and presided over the launch of ASU West. He rode herd on a massive construction program, with thirty buildings going up. But students complained that education didn't get better — class sizes were too large and it was difficult to enroll in required courses.

Now the hunt was on for a new ASU president. The fourteen-member search committee included educators and regents as well as civic and business leaders. Jack, who rarely seemed to get left off a committee, was among them.

As the committee weeded through more than two hundred contenders, one name rose to the top: Lattie Coor, a native Arizonan and president of the University of Vermont.

Coor didn't want his interest made public, and Jack played a lead role in courting him privately. One Sunday afternoon, SRP executive Mike Rappoport was startled to get a long-distance call from Jack. He was in Vermont, and Coor's teenage daughter was worried that she'd have to give up dressage if she came to Arizona. She'd heard that the only riding was Western style. Jack remembered that Rappoport's daughter had ridden horses for years, so he called to check with her. No problem, Coor's daughter was assured.

"This was just so typical of the Pfister approach," Rappoport said.[10] He was the executive of a major corporation, but he was also a problem solver. If he needed some information, he went to the best source.

The search team narrowed the list to three finalists, but two of them withdrew. So it recommended Coor as the sole candidate. Jack praised him as someone "who can lead Arizona State University into the 21st century."[11] The regents unanimously approved the choice in May 1989.

Phoenix attorney Andrew Hurwitz, former chief of staff for Governors Babbitt and Mofford, was a bit awestruck when he was appointed to the Board of Regents in 1988. At 40, he was a good decade younger than most of his fellow regents, who were an intimidatingly successful and accomplished crew. Jack took Hurwitz under his wing.

"He's the only guy I ever met who actually offered useful advice in aphorisms," said Hurwitz, who later became a judge on the Ninth Circuit Court of Appeals.[12]

"One of my passions was increasing minority enrollment at the university," he said. "And Jack decided it was going to be a passion of his — not because of me, but because he decided it was the right thing."

The regents had gotten a disturbing reality check from the 1987 report showing Arizona's rock-bottom rank in minority enrollment. But change didn't come easily to immense and complex institutions. Frustrated at the university presidents' slow response, Hurwitz was all for holding hearings and giving them a thorough grilling.

But Jack pulled him aside and said, "Whatever we measure will improve." It was part of his business philosophy: If you want something to get better, measure it. If a company began counting complaints, for example, employees would quickly realize that the goal was to increase customer satisfaction and begin working toward it.

They needed to figure out achievable but aggressive goals for minority recruitment and then track progress, Jack said. How about aiming to retain those students and increase the number of PhDs and technical degrees among minorities?

In fall 1988, the regents adopted a plan calling for the three universities to raise their minority enrollment and graduation rates by 10 percent a year for the next five years. The presidents could choose their own strategies and count on the regents' full support. But if they missed their annual goals, they would have to come before the board and explain to the public why they failed.

A decade later, Hurwitz saw the success of Jack's strategy: "Remarkably, universities that had struggled to enroll minority students suddenly found qualified applicants, and took the steps necessary to make sure those qualified students graduated."

On a board of directors like the Board of Regents, Jack told Hurwitz, your primary job was to set policy, hire good people to carry it out, and not try to run the operation for them. If you were disappointed, you found better people for the job and moved on.

Jack applied the philosophy when Coor arrived. The regents had hired a premier college president, and now they needed to give him space. If they had any criticism, they needed to speak to him directly, not go public.

"And he was exactly right," Hurwitz said. Coor quickly took charge of ASU in a way that his predecessors hadn't been able to do. "I think it was in large part because people knew that the Board of Regents wasn't willing to go around him. Faculty wasn't going to be able to come to us and undo decisions."

Coor's accomplishments included aggressively recruiting bright students from around the country and turning ASU into a major research university.

Hurwitz was new to the annual tuition-setting decision, which was becoming a colorful and emotional ritual laced with protests, signs, and chants. With university presidents proposing a $156 hike in December 1988, students staged a boisterous rally and march before the regents' meeting in Flagstaff. Jingling their keys as percussion, they shouted, "We hate tuition, yes we do, we hate tuition, how about you?" [13]

Ultimately, the regents scaled the increase back to $84. Hurwitz suggested the figure — prompting some delighted students to dub him

"Hero-witz." Jack, however, favored a larger increase. He recognized the pressure from the Legislature to have students bear more of the cost of their university education. And he never forgot the balance between funding and quality.

What worried Jack was the shift in federal aid. "Studies suggest that the federal government has gotten out of the grant business and into the loan business," he said. Foreseeing an issue that would heat up decades later, Jack was concerned about loading students up with debt.

The state needed to provide more grant money, he argued, because it "reduces the amount students have to borrow to get an education."[14]

Jack left the Board of Regents at the start of 1990. It was just the time the search was on to replace the UofA president, who was leaving. Jack was an obvious candidate. But in November, he wrote a memo saying he was bowing out of consideration.[15] "I decided that I did not want to conclude my professional career with another intensive executive responsibility," he said.

In listing his priorities, Jack said he'd found that teaching and writing ranked higher than administrative responsibilities (Typically modest, he described himself as "reasonably proficient" at the latter.). He was eager to get a better foundation in the humanities, a gap he saw in his engineering and legal education. His plan was to return to school and "pursue a disciplined reading program."

He started off the memo with a quotation from Mahatma Gandhi: "There is more to life than simply increasing its speed."

If Jack ever slowed down, however, no one around him could tell.

Getting to Plan 6

Rep. John Rhodes was beside himself. It was September 1980, and the powerful Arizona congressman was certain that Orme Dam needed to be built. And he was equally certain that Salt River Project had a secret agenda to prevent it.

His conspiracy theory ignored the fact that the project was officially dead. President Carter had insisted on the proposed dam's elimination before he would remove the Central Arizona Project from his "hit list." Now, federal officials were developing alternatives to the controversial dam.

Rhodes went to a briefing and listened to the options with growing exasperation. The list ranged from practical steps — raising the existing Roosevelt Dam by fifty feet — to a grandiose scheme to build more than a billion dollars' worth of levees in the Phoenix area.

Rhodes wasn't buying any of them.

"The whole thing smells like hell to me," he told an *Arizona Republic* reporter.[1]

Jack spoke at the briefing. As usual, his focus was pragmatic. The priority, he said, should be to repair and expand Roosevelt Dam, which would reduce flooding risks in Phoenix.

It was just the thing to set Rhodes off. He accused SRP of "playing its own little game" to get federal dollars, ignoring the bigger picture. He was furious that the Army Corps of Engineers was studying other options instead of getting Orme built. "It's just crazy," he complained.

Jack insisted that SRP hadn't backed away from Orme. His priority right then, however, wasn't building a new dam. He was worried about the old dams. A chilling new set of calculations showed that the maximum floods in central Arizona could be much worse than scientists had figured before. None of the dams on the Salt River was built to withstand that kind deluge. If they blew, the path of destruction through Phoenix would be miles wide. The fix was expensive but effective. Raising Roosevelt Dam, the most upstream of the dams, would protect those below it.

Jack told a reporter that he understood Rhodes' frustration. The Central Arizona Project had been under construction since 1973. Once the canal brought water to Phoenix, there had to be a place to store it until needed. And almost no one disputed the urgency of more flood control.

The Interagency Task Force on Orme Alternatives had been set up in April 1977 with the goal of coming up with a quick solution for storage and flood control. Jack and several other representatives from SRP served on its subcommittees. But in May 1978, the task force gave up.

The government regrouped, and the Central Arizona Water Control Study was launched later that year by the Corps of Engineers and the Bureau of Reclamation (known at the time as the Water and Power Resources Service). Now the game became truly complex.

Gov. Bruce Babbitt established a community advisory committee. A separate Technical Agency Group was created to get input from local, state, and federal agencies. With SRP's quasi-public status, Jack was a member of that group. (Since rival Arizona Public Service was a corporation, its chief executive was in Babbitt's group.)

A consultant, Dr. Marty Rozelle of Dames & Moore, helped lead the overall public involvement program, trying to wrangle consensus out of some 120 stakeholders, including SRP.

The name "Orme Dam" may have disappeared from the discussions, but that was just a technicality. The notion of putting some sort of dam at the confluence of the Salt and Verde rivers was very much alive. With the state's two titans of electricity on opposite sides of the issue, "it turned into a wrestling match between Arizona Public Service and Salt River Project," Babbitt said.[2]

For APS and its allies, including Rhodes, Orme Dam was practically sacred, an indispensable way to store Colorado River water
while also providing flood control.

Jack, though, saw very clearly that times had changed since
the Central Arizona Project was approved in 1968. Orme Dam had
become a magnet for opposition and bad publicity. As Babbitt recalled,
"He, even more deeply than me, saw that we simply had to move
away from just sort of sticking blindly to the past."

Opponents drew on powerful images, including the iconic bald
eagle that would lose nesting areas and the destruction of a large part
of the Fort McDowell Yavapai Nation's reservation along the lower
Verde River. There were "Stop Orme Dam" T-shirts. There was the
David and Goliath tale of a tiny tribe that refused a $33 million offer
to move. There was growing resistance from local fun seekers, who
would lose a favorite spot to ride inner tubes down the river.

Publicly, Jack continued to support a dam at the confluence of
the Verde and Salt rivers. But over the years he'd raised doubts about
the value of Orme Dam. Following the 1978-79 floods, Jack said that
even if Orme had been in place, SRP would still have been forced to
release so much water that crossings over the Salt River would have
been closed.[3] He was also concerned that Orme could cause problems
for the rest of the dam system SRP managed.

Orme had to go. But Jack wasn't going to say it outright. Nor
did he see any reason to tangle with APS. Getting into a fight just
wasn't part of Jack's temperament, Babbitt said.

Instead, he pushed for a process to evaluate the alternatives.
Like a master chess player, Jack saw the course that an opening move
would set in motion. Rational people, working methodically, were
bound to reach the conclusion to eliminate Orme Dam.

"He really wasn't an environmentalist," Babbitt said. "I think
that's the wrong way to characterize him. I see him more as a civic
reformer."

With a handful of other members of the Technical Agency
Group, Jack had a heavy influence on the planning document produced in 1980. It showed *how* to reach a decision: There would be

three phases, each with lots of public comment. It showed *what* might be in the proposal: There was a list of dam expansions, new dams, and reservoirs that could be put together in various combinations.

"Jack Pfister compared this whole process to the Rubik's cube," a hot, new toy at the time, consultant Rozelle said.[4]

But everyone wasn't really willing to play. Many started off with fixed positions, particularly on the confluence dam. "Jack was one of the few who from the outset was very open-minded," Rozelle said. His view was "There's likely to be another way to do this: Let's be creative."

The stroke of brilliance for solving the Rubik's cube came from Wesley Steiner, head of the state Department of Water Resources. He realized that flood control and water storage didn't have to be accomplished by one structure in one spot. The two functions could be handled in different places. Raising Roosevelt Dam would go a long way to reducing flood risks. Lake Pleasant, a reservoir north of Phoenix, could be used to store excess water — the extra room could be created by building a larger version of the dam there, which would be called New Waddell Dam.

Finally, in 1981, the alternatives were refined into seven plans, plus the status quo as a baseline. All the plans, except the status quo, were the same on several basic points: enlarge or replace Roosevelt Dam; put a new storage and flood control structure on the Verde River (called Cliff Dam); and repair Stewart Mountain Dam, the one that came close to collapse the year before.

The big difference among the plans was which storage method they included: a version of Orme Dam (now dubbed the "confluence dam") or New Waddell Dam.

Jack knew he would have to maneuver carefully to back SRP away from Orme Dam. So he immediately formed an in-house task force to analyze the alternatives, focusing the discussion on facts, not emotion.

The task force picked Plan 3, which included the confluence dam, as the best. But Plans 6 and 7, which both called for the New Waddell Dam instead, were SRP's second choice. Now, based on facts and scientific evaluation, Jack had positioned SRP to accept a plan without Orme Dam.

Anyone who wondered about the public relations argument just had to check the news reports as dozens of members of the Fort McDowell Yavapai Nation staged a three-day protest march from their reservation to the state Capitol in the still-blistering September heat.

On October 2, Babbitt's advisory committee voted nineteen-to-one to support Plan 6. The holdout was from the environmental side, Dr. Robert Witzeman, representing the Maricopa Audubon Society, whose objections included Cliff Dam.

Jack endorsed the decision. "This is the plan that has the best chance of implementation," he said. "I'm going to work with the congressional delegation to see that we get it."[5]

Interior Secretary James Watt tentatively approved Plan 6 in November 1981. The final OK would come when the environmental impact statement was completed.

For Jack, the drama was far from over. There were still rear-guard actions to bring back the confluence dam. The worst was an ambush in 1983. Keith Turley, head of APS, suddenly rejected Plan 6. Out of the blue, without a word to anyone else involved in the negotiations, he went straight to the Interior Secretary and announced he was supporting Orme Dam again. Turley said an independent study found that Plan 6 was based on invalid assumptions.

Turley called Jack the day after that startling announcement to tell him about the study. "It was a surprise to me," Jack said in what had to be a huge understatement. Whatever anger he felt, though, he didn't show it in public. He focused on getting the obstacle out of the way.

"The APS report will have to be reviewed by the Bureau of Reclamation and the Army Corps of Engineers to see if they are willing to change their opinion," he said. "They are the ones with the ultimate responsibility of the design of the dams."[6]

Since the two agencies had just put $10 million into examining the options, there was little doubt over whether a quick last-minute study would sway them.

So Plan 6 survived even though the prospects of getting full federal funding seemed dimmer and dimmer. The total cost of Plan 6 was now $1.1 billion, and the mood in Congress was getting chilly. If construction depended on federal dollars alone, the projects wouldn't be finished until well into the next century. So Babbitt set up a cost-sharing committee to drum up local contributions.

The committee, though, was seriously divided. Not everyone cared whether all of Plan 6 got built. Those who worked with the Central Arizona Project, for instance, were only interested in water storage. They didn't want to put any local money into flood control.

"The Plan 6 consensus is unraveling," Jack warned in February 1985. The whole plan had to be kept intact, he argued. He reminded the cost-sharing committee that the danger of catastrophic flooding wasn't hypothetical: "There was a very real possibility that Stewart Mountain Dam could have failed."[7]

Among those who worked to create the Central Arizona Project were Congressmen John Rhodes and Eldon Rudd, Jack Pfister, Sen. Dennis DeConcini and Gov. Bruce Babbitt.

A magic moment interrupted the arguments. On November 15, 1985, water from the Central Arizona Project began flowing to Phoenix. Babbitt and Interior Secretary Don Hodel pushed a button to activate the pumping station for the last stretch of aqueduct into the metro area.

It took until spring 1986 to get a deal on local up-front funding for Plan 6. And then SRP almost derailed it. At the last minute, SRP wanted to be protected from any lawsuits stemming from the plan. Babbitt intervened and held a closed-door meeting between SRP officials and major Valley cities in his office. SRP agreed to accept limited liability for operating the dams.

Jack defended the secrecy. "Your public position is often different from your negotiating position," he said. "Whenever you get a whole bunch of elected officials together, they have to be worried about their public position as well as their negotiating position."[8]

Now the pieces came together. On April 15, 1986, Babbitt and federal officials signed a $371 million cost-sharing agreement for Plan 6: $63 million from Phoenix and five other cities; $81 million from the Maricopa Flood Control District; $52 million from SRP, and $175 million from the Central Arizona Water Conservation District.

One last crisis lay ahead. Conservation groups aggressively opposed Cliff Dam, which jeopardized nesting sites for bald eagles. They had sued to block it in 1985. Now they were stirring up another fight in Congress over the Central Arizona Project. They didn't back off until a deal was reached in 1987. Cliff Dam was scratched. Flood control would be boosted instead by repairs to the two existing dams on the Verde River, Horseshoe and Bartlett. In exchange, environmentalists would not only drop their lawsuit to Plan 6 but also, at long last, end their opposition to the Central Arizona Project.

Jack had sparred with environmentalists throughout the long path to Plan 6. However much they disagreed with him, they respected him. Carolina Butler, an activist and "ordinary housewife" who spent a decade fighting Orme Dam, had two main impressions of Jack: "No. 1, he was always a gentleman. No. 2, he was always honest."[9]

The Audubon Society's Witzeman faced Jack at one point in a television debate. Jack knew how to make his case vigorously but without hostility. Witzeman wondered a bit about how Jack, given his character, had ended up in the constrained role of running a utility. "He was in a sense a Renaissance man, who could see all kinds of perspectives."[10]

New Waddell Dam was finished in 1994, increasing the size of Lake Pleasant, a major recreational spot, by more than ten square miles. Colorado River water is pumped to the Lake Pleasant reservoir in winter, when electric rates are low, and released through a hydropower generating station when demand for water and power rises in summer.

Modifications to Roosevelt Dam were completed in 1996, raising the crest seventy-seven feet to increase storage, providing the additional water conservation space that Orme Dam would have created.

Every year, the Fort McDowell Yavapai Nation celebrates the survival of its reservation with Orme Dam Victory Days.

Jack speaks at the re-dedication of Roosevelt Dam in 1986, the year work began to raise the structure's height to provide additional water storage space.

Jack testified, served on a task force subcommittee, and attended countless meetings over Plan 6. But much of his work was invisible. "His whole modus operandi was to remain out of the limelight," Babbitt said.

What if there had been no Jack Pfister to mediate, keep the focus, and steer the course away from controversial Orme Dam?

"There would have been a knockdown, drag-out public fight between the CAP (Central Arizona Project) establishment on one side, the environmentalists, the Native Americans and the Congress and Administration on the other," said Babbitt. "Now, what the outcome would have been is anybody's guess. The Central Arizona Project was never in danger of being defunded completely. It was too far along by then. But the level of congressional support might well have been scaled way back."

The Crucial Connection

Tom Sands was an engineer who knew how to avoid talking like one. The skill came in handy after he came to Salt River Project in 1974. When he made presentations to senior staff, he'd give them the CliffsNotes version of a program, a dam, or a spillway. "I would simplify it, maybe scale it back, and use descriptions that would cover 90 percent of the issue," he recalled. "You don't want to overburden them with details."[1]

But Jack, with his background in engineering, wasn't satisfied with broad brush strokes. He wanted to know the fine points. "Not only was he a big-picture guy," Sands said, "but he could get down into the weeds and talk the technical details of a particular project."

The two men got along well, and a routine evolved. When Sands was going to make a presentation to the SRP board, he would give it to Jack first. "He would coach me and give me tips and make suggestions," Sands said.

Jack went over the reasoning and thought process that went along with the numbers. Yes, a presentation had to be technically sound, but then Sands also had to think like a company manager and show how projects fit into the overall corporate strategy.

As the Central Arizona Project moved forward in the early 1980s, the SRP board faced a critical decision. Would SRP allow its own gravity-flow system to be used to deliver water from the Central Arizona Project to cities, other farmers, and industrial users — or would it force them to spend hundreds of millions of dollars for a separate delivery system?

Perhaps the biggest issue was water quality, especially for farmers. They had enough trouble already with the variance in mineral content between the Salt and Verde rivers and the poor quality of groundwater that some areas received. The Colorado River had a higher level of total dissolved solids, which would lower the seed germination rates and reduce crop yields. Water from the Colorado also had more chlorides, so adding it to the mix of water carried by the canals would corrode steel irrigation equipment and water heaters, a concern for farmers and city dwellers alike.

In 1983, the Central Arizona Project was just two years away from bringing water to the Phoenix metro area. Although SRP would charge cities for using the canals to carry water to municipal filtration plants, some board members still resisted the idea. "No one can make us deliver their water for them, and I don't think we should offer," one said.[2]

Jack, though, was a realist. He knew he would have to bring the board around to understand that they didn't really have a choice. If SRP didn't make the connection voluntarily, the cities would certainly launch a long and expensive battle to force it to.

Sands was busy calculating the costs of a facility to connect the two water systems and then toting up the effects of adding flows from the Colorado River to local water deliveries. Jack approved hiring an agricultural water-quality expert to provide solid data. The findings were no surprise: some impact on seed germination, corrosion, and crop yields.

Sands made a presentation chart meticulously listing the pros and cons. However, the pro column had fewer items. Jack took a look and shook his head. Make the two lists equal, he said: Combine some negatives and find a way to boost the number of positives. Otherwise, given human nature, the sheer visual image could tip the decision.

Jack wasn't a corporate pit bull, fighting single-mindedly for the interests of his company. To Sands, he seemed to see himself more as a facilitator, an intermediary between SRP's interests and those of the cities, state, and Legislature. "I think it occasionally got him in trouble," Sands said.

Sands finally made his presentation to the board, Jack gave some wrap-up comments, and then the board members started talking among themselves.

As the session went on, Sands said, "You're kind of counting votes based on their discussion, and it was uh-oh, this is headed down."

The question of whether to connect with the Central Arizona Project was tipping to the negative side, and the board was on the verge of taking a motion and a vote. Jack jumped in, Sands recalled, and laid out the reality.

You can cast a vote based strictly on the best interests of your agricultural constituents, Jack told them. But if you vote down this connection, I want you to think — Jack pointed dramatically around the table at the board members — how many of you will be sitting around this table a few years from now? You'll upset the cities so much that they'll push for a change in the voting system.

Changing the method of electing the SRP board was a constant threat. The cities would put up with the one-vote-per-acre method, which gave agricultural users a lopsided advantage over urban ones, as long as it didn't significantly impact their interests. A few years earlier, the voting method had been reviewed by the U.S. Supreme Court and, by a single vote, remained unchanged.

Nearly all of SRP's board members came from farming families. And historically, they had always seen themselves as representing agriculture. Now, as Phoenix urbanized, Jack was showing them that more and more of their constituents were urbanites, and the board must represent them, as well. Persuading the board to connect their water system to the Central Arizona Project's "took a lot of cajoling and a lot of hand holding," Jack recounted in a 2005 oral history for the Central Arizona Project.

But on July 9, 1984, the board finally voted "yes." Jack saw the agreement as one of his greatest accomplishments at SRP. And the benefits, it turned out, worked both ways. When Arizona later went through a drought, the connection allowed SRP to supplement its supplies with water from the Central Arizona Project.

Strength From Diversity

Arizona was among the last states to create a holiday honoring Dr. Martin Luther King Jr. The issue turned into one of the ugliest and most divisive in its history, dragging on from the mid-1980s to 1992.

The state got one black eye after the other, as the holiday failed at the Legislature, at the governor's office, and at the polls. Conventions bailed out, and entertainers canceled appearances. The Super Bowl dumped Arizona as host of the 1993 game. The state was labeled racist, bigoted, backward, the equivalent of South Africa in the apartheid era.

For Jack, it was painful to see his home state repeatedly reject the opportunity to celebrate Dr. King's message of tolerance, equality, and freedom. And he became a force to promote it.

The move to recognize Dr. King with a holiday began not long after the civil rights leader was assassinated in 1968. Illinois led the way in 1973 as the first state to adopt an MLK Day. Congress created the federal holiday in 1983.

Arizona didn't have a large African-American population pushing the holiday — less than 3 percent of the population was black in the 1980 census. But the state was poised in May 1986 to create its own King Day. The state Senate had passed legislation that added the holiday and offset the impact by combining Washington's and Lincoln's birthdays into a single Presidents Day. The bill failed in the House by one vote.

Gov. Bruce Babbitt took the issue into his own hands later that month. In a theatrical flourish, he signed an executive order creating Arizona's first Martin Luther King Jr. Day during worship services at an iconic black institution, First Institutional Baptist Church in downtown Phoenix. The congregation clasped hands and broke into the civil rights anthem, *We Shall Overcome*.[1]

The holiday applied only to the offices under Babbitt's own authority, the executive branch part of government. And it was doomed when archconservative Evan Mecham won the governor's race in the upset victory of November 1986.

When the governor-elect listed the changes he'd make after taking office, abolishing King Day was at the top. Mecham argued that he didn't have a choice: The state attorney general, who was threatening court action to void Babbitt's order, had issued an opinion that Babbitt exceeded his authority. (Babbitt, a former attorney general himself, maintained that his order was completely legal.)

But Mecham's motivation went beyond legal details. The slain civil rights leader, he declared, "was not of a stature to deserve a holiday."[2] When a grassroots group met with Mecham, urging him to change his mind, he said that black people needed jobs, not a holiday.

Stevie Wonder gave Arizona a hint of the potential damage to its image and economy. Wrapping up a concert in Tucson in November, the legendary musician warned that he wouldn't play in Arizona again if Mecham canceled the holiday.[3]

The issue was escalating far beyond the question of whether to give state government employees a paid day off. Now the debate was also about Arizona's identity and its future. Was this place really so intolerant? Would the state miss the chance to use its changing demographics as an asset, not a liability? That Western streak of independence was also in play, with many Arizonans bridling at pressure from outsiders.

As president of the Board of Regents, Jack tried to steer clear of the controversy leading up to Mecham's inauguration. He was someone who played by the rules, by temperament and legal training — and if there was a problem with the rules, he pushed to change them, not flout them. The attorney general's opinion had put a spotlight on the limits of acting independently of the Legislature.

But the regents had to face the issue at their December 1986 meeting. A group from Arizona State University had presented them with a petition, signed by more than 1,200 students, faculty members, and staff workers, seeking a holiday to honor Dr. King. ASU student Hedy Jacobowitz recalled how she felt the previous January, when many school districts and community colleges took off a day to honor King: "embarrassed and infuriated that the Board of Regents failed to take the same action."[4]

But Jack responded that the board didn't have the authority to declare a holiday for university workers, since they were actually state employees. The board had already supported less formal steps, passing a resolution the previous January asking the three campuses to hold observances for Dr. King.

Within days of taking office, Mecham pulled the plug on King Day. At the time, only eleven other states didn't officially celebrate the holiday. None had rescinded it. With one signature, Mecham turned Arizona into a pariah.

By mid-May, the Phoenix area had lost twenty-eight conventions, with an economic impact of $18 million. Yet King's legacy was being saluted on the local level. At least seven Arizona cities — including Flagstaff, Phoenix, and Tucson — had made King Day a paid holiday for municipal employees.

Despite a barrage of pleas and threats, the Legislature stalled on the issue in 1987. But it pushed Mecham out the door in April 1988, removing him from office over misuse of funds and obstruction of justice.

Meanwhile, the state attorney general had issued an opinion saying the regents had the power to adopt holidays for university employees. They voted in September 1988 to establish a paid Martin Luther King Jr. holiday at universities, preventing any added cost to the state by requiring each school to eliminate an existing paid holiday.[5]

The new governor, Rose Mofford, hoped the regents' decision would prod the Legislature into action. But while the House approved the holiday in 1989, the bill died in the Senate.

Yet the urgency was growing as the Phoenix area angled for the 1993 Super Bowl. Bill Shover, director of public affairs for Phoenix Newspapers Inc. and a longtime friend of Jack's, was chairing the drive. That meant he was also pushing for an MLK holiday, because without it, the big game was never coming to Arizona.

Jack joined an all-star list of business and community leaders pressing the Legislature to go into a special fall session in 1989. By now, Arizona was one of just four states that didn't commemorate Dr. King. "It is a sizable negative," Jack pointed out, picking non-confrontational language designed to sway legislators, "particularly at a time when we're trying to attract new industry to the state."[6]

On September 21, lawmakers finally approved King Day. But the celebrations didn't last much longer than champagne bubbles. Italian-Americans were furious that legislators offset the new holiday by eliminating Columbus Day. King Day opponents capitalized on the anger and quickly collected enough signatures to put the question on the November 1990 ballot.

But Phoenix did land the Super Bowl in March, a surprise victory that left sports fan and businesses giddy. There was one condition: Arizona had to have a Martin Luther Jr. King holiday. So the Legislature scrambled to end the uncertainty by passing *another* version that restored Columbus Day and created a combined Presidents Day. It superseded the previous holiday and made the referendum moot.

The drama wasn't over. King Day foes, this time reinforced by allies of former Governor Mecham, launched *another* petition drive and put the new version of the holiday on the ballot. But in a crazy legal twist, that put the first referendum back in play.

The two variations of King Day split the vote, so both ballot measures failed even though a post-election poll found that 63 percent of voters said "yes" to at least one of them. As an added complication, a news report just two days before the November 6 election said Phoenix would lose the Super Bowl if Arizonans failed to approve King Day. Thousands of Arizonans felt they were being strong-armed and voted "no" in defiance.

The day after the election, the commissioner of the National Football League recommended moving the 1993 Super Bowl out of Phoenix. The boycotts started again, even worse than before.

Arizonans were scrambling to respond. Dr. Warren H. Stewart Sr., pastor of the First Institutional Baptist Church, hadn't played much of a role in the campaign for the dual ballot propositions. Now, friends and allies urged him to help lead a new campaign.

Within days of the election, he held a gathering to organize the next push to honor King's legacy. The campaign would be called Victory Together. Jack came to that first meeting and asked to be part of the effort. "He said, 'I'm not trying to lead this,'" Stewart recalled. "I was impressed that a man of his stature in the community could be a follower and a partner."[7]

Jack helped Stewart connect with the powerful members of the Phoenix 40 and overcome their doubts about a Baptist minister known as a firebrand. "They weren't convinced that a coalition led by me would be able to win the holiday," Stewart said. "Jack was instrumental in saying: 'Work with this guy.'"

Once again, the Super Bowl was the catalyst for the Legislature to act. In March, five days ahead of a meeting of National Football League owners, lawmakers voted to put a proposal for a paid state King holiday on the November 1992 ballot. The NFL still bounced the 1993 game out of Phoenix, but it awarded the 1996 Super Bowl to the city.

Now the challenge was to get a "yes" vote. Different states had such varying versions of the holiday that it was hard to make comparisons, but by some tallies, Arizona was the sole holdout. Nowhere, though, had a holiday gotten approval at the polls. "We do have the opportunity to be the first, and only, state to pass a holiday by vote of the people," Stewart said.[8]

Jack did everything he could, in his typical behind-the-scenes style, to make sure that happened. He went to virtually every meeting of Victory Together. He helped the group set up interviews with political consultants and evaluate them — a vital part of the path to success, Stewart said.

Steve Roman, senior vice president at Valley National Bank, had worked on the 1990 campaign and was among the leaders in the new one. He met almost daily with Jack, the consultants, and Victory Together's statewide director, the Reverend Paul Eppinger, to hash out the gritty details of the campaign. Jack was on the steering committee, the finance committee, and the legal-advisory subcommittee, which he co-chaired.

With his corporate contacts, he helped the group put together a war chest that reached a million dollars. In contrast to the rallies, speeches, and marches of the 1990 campaign, Victory Together deliberately remained low key, concentrating on voter registration to expand the pool of likely "yes" votes.

"It was a very carefully designed strategy to develop a broad, grass-roots coalition of support and to do that through various networks, and in having publicity only at the very end," Jack said.[9]

The holiday succeeded with an overwhelming margin, getting 61 percent of the vote.

The next week, supporters of King Day held a huge luncheon celebration in the ballroom of a Phoenix hotel. Reverend Eppinger was the master of ceremonies, and he jokingly recounted how he became certain the night before the election that this time the holiday would pass.[10]

He'd been at a prayer service at the First Institutional Baptist Church, and he spotted Jack Pfister and Bill Shover in the pews. As the choir sang with passionate commitment, Eppinger told the lunch crowd, Jack and Bill began dancing and singing in the aisle in anticipation of winning. When he saw that, Eppinger continued, "I knew we had the victory."

Of course, the story wasn't true. The two men *had* been at the service, but they were the last people you'd expect to see capering about anywhere. The audience burst into laughter, and no one was more amused than Jack.

Arizona was no longer an outlier. Over the years, at least one hundred sixty-five groups and unknown numbers of tourists had boycotted the state over the King holiday, costing the state an esti-mated $190 million.[11, 12]

Jack is among the handful of people who are the reason Arizona now has a King holiday, in Roman's view. "It could not have happened without him," he said.[13]

In 2006, Jack received the Martin Luther King Jr. Servant Leadership Award at Arizona State University's annual MLK celebra-tion breakfast. An article about the honor called him "a driving force behind the passage of an MLK holiday in Arizona."[14]

No Hat in the Ring

"Pfister for governor." Over the years, Jack heard from many people who wanted to put that bumper sticker on their cars.

The suggestion picked up real momentum during the drive to get rid of Gov. Evan Mecham. After winning in a fluke, three-way race, Mecham had become a one-man wrecking squad for the state's image and economy.

Even before the recall drive started in July 1987, the moment it was legally permitted, a guessing game had started: Who would run to replace him? Jack's name was "thick in the rumor mill," a political item in the *Arizona Republic* noted on August 15, even though he waved off the possibility. He didn't really need the state's top job, the article went on to say: "Some have speculated that Pfister, 53, already has a lot of the influence and power that comes with being governor."

Jack loved talking politics. But he was more of a tourist. He didn't want to live in the divisive, polarized world of elected government. Pat, he knew, would hate the relentless spotlight of being the governor's wife.

Gov. Rose Mofford and Jack Pfister

Still, friends and colleagues urged Jack to jump into the race —
he could be the healing force that Arizona desperately needed. James
G. Derouin, an attorney who had been representing SRP, made an
eloquent appeal in a P.S. to a letter about another topic. "You should
seriously consider being the consensus recall candidate that the establishment is seeking."[1]

Derouin's note pushed all the buttons he knew would appeal
to Jack: duty, a noble purpose, a good cause, and saving "the soul of
your home state."

In September, Jack was one of eight potential candidates who
received a letter from the Mecham Recall Committee asking them
to refrain from announcing their candidacy until the success of the
recall was assured.

The issue turned out to be moot. Mecham was impeached, convicted, and removed from office, and the Arizona Supreme Court ruled
that the constitutional order of succession trumped a recall election.
Democrat Rose Mofford, the secretary of state, became governor.

Getting Arizona in Harmony

As he campaigned for the Martin Luther King Jr. holiday, Jack recognized the issue went much deeper. The state had to deal with the underlying problems of racism and discrimination.

In October 1990, he'd joined a nine-member panel in a live television broadcast on the local NBC affiliate, to discuss "Racism in Arizona: It's Getting Worse. We Need Solutions." From chatting with his fellow panelists, he knew there was a hunger for an organization to tackle the challenge.

As always, Jack attacked the issue by educating himself. Searching for the best way to build bridges between diverse communities, he found the right framework in a Kansas City organization called Harmony in a World of Difference. It had been founded in 1989 after a survey ranked Kansas City tenth among the nation's most segregated cities. The group's goal was to promote mutual respect among people of different backgrounds, and its strategies included classroom materials, cultural-awareness workshops, and task forces in key areas.

Jack found the perfect partner in an energetic African-American woman who was relatively new to Phoenix. Loretta Avent had moved to Phoenix from Washington, D.C., at the end of 1988 when her husband, Jacques, left his job at the National League of Cities to take an executive position with the Phoenix City Council. Loretta, a well-connected consultant and long-time friend of Bill and Hillary Clinton, had met Jack shortly before moving to Phoenix, and the two immediately clicked. They shared the same commitment to community, respect for individuals, and upbeat conviction that reasonable people could find solutions to thorny problems.

Temperamentally, though, they were virtual opposites. Loretta was effusive, outspoken, voluble, dramatic — the kind of person described as a "live wire" in feature profiles. Jack was deliberate and purposefully low key. "It was like yin and yang," said Ioanna Morfessis, who was president and CEO of the Greater Phoenix Economic Council at the time.[1]

These were tough times for Arizona. "Foreclosed" signs dotted the streets as the state struggled to recover from the latest bust in the real estate market.

In 1991, Jack wrote a New Year's Day opinion piece for the *Scottsdale Progress* with an icily candid analysis of Arizona's political and economic condition that would have sobered up the most carefree reveler. Indeed, the state had spent the 1980s like a reckless party animal, indulging in economic and political excesses, in Jack's view. Investments were based "more on an unrestrained optimism in Arizona's real estate than on the intrinsic value of the investment." The political climate had become confrontational. And Arizona suffered from an international reputation as racist and bigoted.

"We can debate the merits of this perception or how we compare with other areas," Jack wrote, "but most minorities will tell you that Arizona often presents an inhospitable environment. Examples of racism and discrimination abound."

For the state to become internationally competitive, it had to welcome ethnic and cultural minorities and provide the educational and employment opportunities for them to succeed. Jack began laying the groundwork for the new organization that would pursue those goals. For legal reasons, Arizona couldn't adopt the Kansas City name, so the group became Harmony Alliance Inc.

Jack turned to Loretta Avent to help lead it as state coordinator. She agreed, with three conditions. Jack had to chair the group. She didn't want a salary. The meetings would start at 7:30 p.m. and end promptly at 9, because, in her experience "when you go too long, it's trouble."[2]

The group would meet in places of worship, a different one every time, emphasizing its message of outreach and respect.

Jack began building a representative slice of Arizona's minorities: racial, ethnic, cultural, and religious. He didn't send out a general letter of invitation, he spoke with each person individually. The personal touch, despite all the other commitments in Jack's life, made a huge impact. "When Jack Pfister called you and said, 'Do you have a few moments, I'd like to come by and visit this week?' you said yes immediately," Morfessis said. "Everyone knew his agenda was for the common good, and I think that's why people listened to him so intently and responded."

People also knew an invitation from Jack was a commitment to actively participate. "Jack didn't let you off the hook once you said yes," said Morfessis, who found herself serving on various Harmony Alliance committees after joining the board.

In the end, Jack assembled a rainbow board of directors that included members who were Asian, Hispanic, black, Muslim, and Jewish. There were CEOs of major corporations, church leaders, and community activists representing most ethnic minorities in the state.

The group was up and running by March 1991, but it flew under the media radar while gaining momentum. In the next five months, Harmony would recruit more than 100 members.

It adopted a mission statement: "to promote harmony in the state of Arizona and the full integration of all segments of our diverse population into the economic, educational, political and power structure of the state."

Harmony Alliance was truly breaking new ground. Arizona had interfaith groups that fostered understanding among different religions. The nonprofit Arizona Humanities Council promoted a broad range of cultural activities. Local governments, such as Phoenix, had human relations offices to promote tolerance. Ethnic festivals showed off individual groups' music, crafts, and food. But no organization had brought together Arizonans from so many different backgrounds and beliefs — and got them working together on long-term projects.

Jack set a warm tone at meetings. "We all felt equals," said Rabbi Robert Kravitz, a board member. "We called him 'Jack,' we didn't call him 'Mr. Pfister.' He was a buddy and a friend and always listening."[3]

Jack focused on the spheres he knew, particularly business, where he saw enormous potential for both improvement and payoff. Some suspected that companies were merely being self-serving when they tried to be inclusive of minorities. But Jack argued a rising-tide theory: As the business community became more welcoming, all boats would float higher.

"He was an inspired leader," Kravitz said. "The corporate world, as I viewed it at that time, wasn't quite as open to diversity as Jack was. He was maybe two steps ahead, and leading this whole gang of us with him."

Jack poured time, work, and money into getting Harmony Alliance going. "I remember him pulling together a lot of leaders and introducing them to us, often doing it over lunch and dinner, and he picked up the tab each time" said Lisa Loo, a board member and attorney. "For a number of events, he personally wrote the check."[4]

In the summer of 1992, many Arizonans saw a state that was still a long way from achieving equality, although minorities perceived the problem as much worse than Anglos. A June poll by the *Arizona Republic* found that 89 percent of minorities and 78 percent of Anglos believed the state had at least some degree of racism.

Just 40 percent of minorities thought job opportunities were equal for all. The prejudice wasn't necessarily blatant. "There's something you can feel in the atmosphere," said one Hispanic woman who sensed the deck was stacked against her at job interviews, "and you look around and there are no minorities anywhere in the office. That's a clue."[5]

Harmony's public debut didn't come until October 1991, when it issued a press release introducing the new grassroots organization. "We want every conceivable cultural viewpoint represented," Jack said in the release. "To be successful, Harmony must be inclusive and not exclusive."

He outlined plans for educational outreach, workshops, round-tables, and a Phoenix festival the following October to celebrate cultural diversity.

Arizona wasn't worse than other states, Jack told a reporter for the *Scottsdale Progress* in a November article. "But still it is a racist state."[6] He went on to give an example: "I'm told that black children in the Scottsdale schools don't find it a comfortable place to be." His frank comments created an uproar.

The Scottsdale school district superintendent wrote a letter to Jack expressing "consternation" and describing the district's commitment to cultural diversity.[7]

Jack responded cordially, explaining that his comments were based on a discussion with a black parent with children in the district. He welcomed the superintendent's offer of a visit to the schools. Spotting a chance to make lemonade from a lemon, he also suggested they might work on a project for in-service training for teachers.

"Since I was interviewed for the *Scottsdale Progress*, I used the Scottsdale District as an example," he explained in his letter. "The problem exists in most school districts. It is one of the reasons why the Harmony Alliance was formed."[8]

Arizona certainly needed something to counteract what a November 6, 1991, editorial in the East Valley *Tribune* newspapers described as "ethnic friction that ranges from discomfort to xenophobia to full-blown racism." In a particularly high-profile case, racial overtones in the death of a black man in a supermarket parking lot had triggered vigils and demonstrations the previous summer. Passions were further inflamed when an all-white jury acquitted the two store employees, white and Hispanic, accused of crushing his larynx in a violent confrontation. The editorial also cited attacks on a Chinese Christian Church and the continuing discrimination that Hispanics face. "At the very least," it opined, "the newly formed Harmony Alliance is chicken soup. It can't hurt, and it might speed the cure."

Roundtables, where people from a particular industry or area of interest could hash out ideas, were one of Harmony's major remedies. They would analyze the strategies at organizations that were "best in class" for diversity, figure out how to use them locally, and then set measurable goals.

"Jack believed in benchmarks," said Rabbi Kravitz. "There had to be metrics, a way of tabulating what was going on in real numbers and not just assumptions."

The reports from the roundtables would be distributed in the community, publicizing the benchmarks. Harmony's focus, in Jack's typical fashion, was in moving forward: "Rather than criticize organizations for past poor practices we will celebrate outstanding practices as a way of offering hope for the future."

For a new group that started with a bank account of zero, Harmony had ambitious plans. Besides the roundtables, it planned two back-to-back events in Phoenix at the start of October 1992: a conference for businesses called "The Dividends of Diversity" and a festival in Phoenix.

This was an organization with no paid staff that relied on the voluntary efforts of members whose calendars were packed solid. It looked for cash and in-kind donations from corporations.

As the actual events got closer and closer, Jack was starting to worry about whether he could raise enough money to cover them. So in mid-September, he sent a $10,000 personal check to the Arizona Community Foundation, Harmony's financial agent, to be held in reserve in case of a shortfall. But Jack did such a successful job of fundraising that the check never had to be cashed.

Some seventy-five people attended the Dividends of Diversity conference on Friday, October 2, to network, discuss issues, and hear speakers with expertise in pluralism and employee relations. Harmony also held its first roundtable that day, bringing together managers from the public sector.

Saturday's Harmony Festival was an all-day extravaganza, lasting from 10 a.m. to 6:30 p.m. The idea wasn't totally without precedent. In 1981, Phoenix Arts Coming Together had sponsored a festival called "Hello, Phoenix!" that became a celebration of the city's ethnic roots.

But Harmony covered the full spectrum of the Valley of the Sun. Brazilian, Thai, Native American, Turkish, Mexican, Norwegian, Jewish (after sundown, because it was the Sabbath) — it was a glorious picture of the region's cultural assets. There was no admission charge, and the "ethnic treats" were priced at a dollar or less, so people could afford to sample lots of different cuisines. The day started with a parade, offered crafts and storytelling for kids, and featured three stages with nearly forty different performances of music and dancing.

An estimated ten- to twelve thousand people attended, a respectable crowd considering there were other events in downtown Phoenix that day, including an anti-crime program and Oktoberfest. What made Harmony's celebration special, said attorney Loo, was how engaged the crowd was. "It wasn't one of those festivals where people just kind of walked by and admired the booths from a distance. This was one where the people who showed up were really interested in the organizations that participated."

Harmony optimistically called this "the first annual" festival. But it was the only one. As a brief history of the group later explained, "The Festival was a wonderful success, but for future reference, it should be noted that funding is very difficult to obtain, and the work required to produce such an event is basically impossible without a full-time staff, regardless of the number of volunteers who might be willing to lend a hand."[9]

Harmony Alliance sponsored a media roundtable, which panelist Richard de Uriarte described in a follow-up column in the *Phoenix Gazette*. He'd heard from readers about subtle signs of bias and stereotypes. When a crime occurred in the heavily Hispanic western part of Phoenix, one complained, the headline always seemed to mention the "west side" — while crimes in other places weren't branded with the location. Other panelists talked about de facto segregation in radio, as stations sorted themselves by format, and the importance of avoiding stereotypical scenes of mariachis and folkloric dancing in TV segments about the Hispanic community.

The panel was just the sort of reality check that Jack had envisioned when he founded Harmony Alliance.

Not every project got off the ground. Plans for a TV program called *Phoenix Rising,* to be produced through Scottsdale Community College, fell through when the person who had proposed it no longer had the time to volunteer.[10]

One of Harmony's hottest topics was how to deal with country clubs. The two local powerhouses, the Phoenix and Paradise Valley country clubs, denied that they discriminated, but their membership was all white and gentile.

In spring 1992, Harmony Alliance swung into action.[11] The reception for the new president of the Phoenix Union School District, a Hispanic, was scheduled to be held at the Phoenix Country Club. After its March 4 meeting, Harmony decided it was time to send a message. Members went straight to their phones and made some strategic calls urging the district to pick another site. The next day, the reception was moved to a different location.

Arizona Town Hall, a statewide forum, had held meetings for some twenty years at the Paradise Valley Country Club. After Harmony raised concerns, the group decided to go somewhere else for its upcoming training session on diversity.

Harmony Alliance, which counted about four hundred members by then, flirted with organizing an outright boycott of clubs that failed to promote diversity. That was the tactic local civil rights activists, led by African-American entrepreneur Lincoln Ragsdale, were threatening to use. But in the end, Harmony opted to use encouragement rather than muscle.

"My sense is that there's a genuine interest within the business community in seeing that the exclusive clubs have diverse members," Jack said. In reality, as he pointed out to a reporter, "all we have is moral persuasion" to bring about change.[12]

Persuasion paid off. Harmony's efforts were certainly one reason the board of the ninety-year-old Phoenix Country Club voted to admit two African-American couples in September 1992. Minority leaders said these were the first black members of a country club in the entire metropolitan area.[13]

◇ ❯ ❯

Throughout the second half of 1992, Loretta Avent was juggling Harmony with another project close to her heart: getting Bill Clinton elected president. When Clinton took office the following January, she was off to Washington, D.C., to become special assistant to the president for intergovernmental affairs and liaison to the First Lady's staff.

Harmony lost one of its main engines when Avent left. The board tried to develop an activity plan for 1994, but the momentum was gone for a variety of reasons.

"The immediacy of the need had passed," said José Cárdenas, an attorney who served on the Harmony board.[14] Arizona finally had its King Day. The award of the 1996 Super Bowl was a signal that the state was no longer an outcast. After years of fighting so hard for the holiday, many supporters had battle fatigue. Meanwhile, a variety of other diversity projects was under way.

Although a lot remained to be done to heal the divisions in Arizona, there was a feeling that Harmony had served its purpose.

"A number of us felt that we accomplished the goals Jack had envisioned," Loo said. Chief among them was presenting a different face of Arizona to the nation. Harmony created lasting connections among people from different backgrounds who might not otherwise have met. It spread ideas for making diversity an asset, not a problem.

Was Jack wearing rose-colored glasses in thinking that reasonable discussions could patch up the rips in Arizona's social fabric? "I think there were people who thought he was a Pollyanna, but I never saw that myself," Loo said. "They may have thought his idea to reach consensus when possible was contrary to their leadership style."

In 1996, the board voted to dissolve Harmony Alliance and distribute its assets to community groups pursuing its vision.

And yet in many ways Jack's grand idea is as timely as ever. "If the Harmony Alliance recreated itself today, it would still be needed," Rabbi Kravitz said. "The idea that people of all kinds of backgrounds can get together and talk about issues for the general good, for the common good, is a vehicle we don't have today."

Adopted Grandkids

When Jack and Pat Pfister needed a "grandchildren fix," their daughter would get on the phone to Loretta Avent with an urgent message: Get Brittany and Bryant over there.

The little girl and boy were actually Loretta's grandkids. But she and her husband, Jacques, were raising the children, so they had the parenting role. Jack and Pat were happy to take on the grandparenting one. "Jack was somebody they could talk to," Loretta said.[1]

The children can't remember a time when they didn't know the Pfisters, who met them at the end of 1989, when Brittany was turning two and Bryant had just been born. When their parents split up a few years later, the children went to live with the Avents.

Jack became close to the Avent family while he and Loretta were running Harmony Alliance in the early 1990s. Sometimes she'd pick up the children from day care and bring them into the office.

Jack and Pat took to the spoiling part of grandparenting, "They were generous," Brittany said, "especially for birthdays and Christmas." The gifts were invariably checks, so they didn't have to worry about choosing the wrong thing.

The "adopted" grandchildren took the pressure off Suzanne and Scott, who had no plans to provide any.

The Pfisters would try to make all the children's events, whether it was a preschool graduation, Bryant's performance with the Phoenix Boys' Choir, or their many basketball games. "I felt like they were relatives," Bryant said.

Right after Bill Clinton was elected president, the two little children were at Mayor Terry Goddard's Christmas party. There was even a pony ride, and they were waiting to go on when they were called into the house to talk on the phone.

Leaving Bryant to hold their place, Brittany went inside to find President-elect Clinton on the line, with his wife, Hillary. They were calling to explain why her real grandmother wasn't there: Loretta, a close friend of the Clintons, was in Arkansas working on preparations for him to take office. Through the window, though, Brittany saw it was her turn to go on the pony. She dropped the phone and ran out.

Jack marveled at the little girl who hung up on the president, and kidded about it for years.

Jack saw even more of the children than Pat did, sometimes meeting them for breakfast. He loved hearing stories and always wanted to know what they were doing. He encouraged Brittany's idea to start a "grand-families" club so children like her would realize their situation wasn't unique. She finally got a chance to form the organization in 2013, when she transferred to Arizona State University as a junior after community college. Jack didn't live to see its success — he would have been delighted at its recognition as a trendsetting student group.

A dose of the grandchildren would always make Jack laugh, Avent recalled. "He said Brittany and Bryant made him feel young."

Peak Performance

A leader starts off feeling on top of the world. And then things fall apart. The disappointment is entirely predictable, according to Jack's analysis of leadership.

He believed in what one of the people he mentored called "the Pfister Performance Curve."[1] Here's what it looks like, as Jack would map it out on a graph:

The initial high: For the first few months in a new leadership position, people feel effective. They take the easy steps to accomplish their goals.

The dropoff: After that initial honeymoon, performance slides steeply. Leaders realize how much there is to learn. They're at the bottom of the curve.

Climbing up: Leaders become more effective, gaining knowledge, building relationships, and having more insight into the job.

The peak: For the next one to seven years, leaders reach a continuing level of high performance. But it's unusual to be effective much longer.

The slide: The surrounding environment and the goals of an organization change. New leadership approaches are needed, but the power of habit makes it hard to change. Frequently, people are unaware of their inability to adapt, which in some cases may lead to an untimely exit.

In the 1980s, Jack was at his peak at Salt River Project. The high point, in fact, lasted longer than his curve predicted. He planned to finish his career by stepping down slowly and deliberately. But the end was more abrupt than he expected.

As the decade started, Jack had gained confidence leading a giant utility. Now he could start turning management theory into practice. His priority was productivity.

With inflation driving up costs, Jack knew that he had to take maximum advantage of current staff and resources. He picked a ten-member team, which he chaired personally, to begin work on improving productivity in 1980.

By 1982, SRP had developed a set of programs to boost productivity and promote the participative management that Jack saw as its companion.

There were Quality Circles. There was a New Supervisors Institute to instill good communication and management habits from the start. At the next level, there was the Executive Resource Planning Program and the Executive Management Institute.

"He was fascinated with improving managerial acumen and practice throughout SRP," said Mark Bonsall, a rising manager at the time. "He realized that was going to take time and the passage of generations. But he got it started."[2]

Bonsall went through the supervisory training classes, which were taught by a mix of internal staff and consultants. The curriculum was intensely practical, with role playing on issues like performance reviews. Afterwards, the class would go over the videotapes. The experience was "brutal," he remembered, including the embarrassment of watching themselves when they knew they'd fumbled.

But it paid off. "I made mistakes in the lab as opposed to in the field," said Bonsall, who became general manger in 2011. "SRP has reaped benefits for decades."

Jack's foundation program for the entire staff was PEP, the Project Effectiveness Program, which emphasized letting employees identify problems and then figure out improvements. A videotape introduction to PEP used the decline of the privately owned railroad industry in America as a cautionary tale. The video blamed the railroad's fate on complacency and insensitivity to customer needs (an analysis that historians and economists might dispute).[3]

"Employees are expected to become involved, take risks with their recommendations, be responsible, and be committed to following through when ideas are implemented," Jack said. "The expectations are even greater for management. Participative managers are open to competent influence from their employees. They must be willing to allow the rationale for their decisions to be subject to change."

For his own executive staff, Jack brought in authors and speakers on topics ranging from crime to religion to philanthropy. "They would have nothing to do with the electrical industry or water," observed C.A. Howlett, who was SRP's head of special projects.[4] Two professors from Arizona State University once talked about "Humor in interpersonal business relationships."[5] The sessions wrapped up with dinner, for a chance to discuss the topic less formally.

The executive team got almost a classic liberal arts education, Howlett said, although not without some eye-rolling by those who thought they were wasting their time.

"In addition to getting us to think differently, it was a team building experience," said Mike Rappoport, who handled government relations.[6] One point of solidarity was grumbling over the whole exercise — although Rappoport valued the discussions so much that "I made a point of not missing them" despite a heavy travel schedule.

One assignment was the popular *Zen and the Art of Motorcycle Maintenance* by Robert M. Pirsig, a philosophical novel exploring the meaning of quality. Not long afterwards, Pat held a birthday party for Jack in their backyard and invited his staff.

Here was the perfect chance to tweak their boss. As a gag, they all chipped in to buy "this wreck of a motorcycle, leaky and smelling," Rappoport said. They stashed it in the alley behind the house. In the middle of the festivities, Rappoport and another executive rolled it into the yard.

"I was looking at Jack and he was looking at me," Rappoport recounted, "and I thought, uh-oh, we may have gone too far." Then Jack broke into a roar of laughter.

"Jack's view was we have to start thinking out of the box,"

Rappoport said. "He was attempting to alter the culture, at least among his executives, so it was not old school."

Jack had a knack for getting results out of meetings. His approach was very methodical, Rappoport said, and he liked to use a flipboard to write down key points. Everyone would have a say, and Jack would be looking for the points of consensus and writing them down. At the end, he'd make assignments, going through the pages and putting someone's initials next to each item that had to be accomplished. He'd rip off the pages and later transcribe them into a notebook, so he could be sure to follow up.

Jack also had a gift for spotting unexpected paths to solutions. One time he had breakfast with members of the Cancer Society who described how hard it was for children with cancer to be stuck in the desert heat all summer. They'd done some tenting excursions. Now they wanted to buy a camp in the cool, northern pines. They knew the issue would touch Jack's heart.

"The easiest route was to open up his checkbook and write a check," said Ernest Calderón, an attorney and former university regent who was mentored by Jack.[7] But Jack came up with a more effective plan.

By coincidence, he had breakfast the next day with the Boy Scouts. They were trying to raise money to fix up their R-C Camp.

Jack spotted an opportunity. The scouts only used the camp from June to mid-July. Why didn't the Cancer Society take the second half of the summer? Instead of buying a camp, it could pay for improvements and accessibility features at the R-C.

Jack even came up with the idea of reversible signage so the camp could change identity, Calderón said. In 1985, Arizona Camp Sunrise opened at the scout site. With medical staff on call around the clock, children with a life-threatening disease could enjoy a "normal" summer camp experience, with activities like archery, horseback riding, hiking, fishing, and campfires.

"What an ingenious thing to do," Calderón said.

◇◇◇

While Jack never ran for public office, SRP employees did — and their successes included seats in the Arizona Legislature. Art Hamilton, who worked in public relations, was elected in 1972. A Democrat, he served a quarter century, most of it as House minority leader. Republican Chris Herstam joined him in the House a decade later. An SRP land agent at the time, Herstam vividly recalls how Jack called him at home the day after the election. After offering congratulations, Jack said, "Chris, I just want you to know that SRP will never lobby you on any issue before the Legislature. We want you to be an independent thinker and never beholden to your employer."[8]

Shaking hands with Art Hamilton

In his eight years at the Legislature, Herstam declared a conflict on any bills affecting SRP and abstained from voting on them. With the backstop of Jack's ethical standards, he said, he was never accused of a conflict of interest.

In February 1987, Herstam got a dramatic call from Jack. The legislator had crossed Gov. Evan Mecham, who was demanding that Jack rein him in or fire him. "My knees began to shake a bit," Herstam recalled. But Jack made it clear that he had no intention of yielding and just wanted to put him on guard. "Pfister urged me to continue doing what I believed was best for the state," Herstam said. "His vote of confidence was reassuring."

For all Jack's jousting with the SRP board over the years, he had a close relationship with board president Karl Abel. "Karl was really a prince of a guy," Jack said.[9]

He had the qualities Jack admired: reliability, excellent leadership skills, and a warm relationship with those who worked with him. He strongly supported management. And he shared Jack's belief that SRP needed to play an active role in the community.

But in 1982, Abel retired after a decade in office, and the wind began shifting against Jack. SRP property owners elected the logical successor: John Lassen, who had been vice president.

SRP's peculiar status as a half-public entity showed up in the changeover. Lassen took an actual oath of office, and he was sworn in by the chief justice of the Arizona Supreme Court.

He claimed to share Abel's perspective.

But over time, it became clear that Lassen didn't have the same impulse to serve and participate in the wider community. He wanted SRP to stick to basics.

Meanwhile, Jack continued to step up his own and SRP's civic roles.

He served on the boards of directors of local companies, such as United Bancorp of Arizona, Southwest Forest Industries and Del E. Webb Corp.

At various times in the 1980s, he was on the board of a string of organizations, including the Electric Power Research Institute, Arizona State University Foundation, Arizona Humanities Council, Greater Phoenix Affordable Health Care Foundation, Arizona-Mexico Commission, American Red Cross, YMCA, and Museum of Northern Arizona.

Often he served a term as board president. He chaired the Arizona Academy of Public Affairs, which sponsored town halls twice a year.

He was on countless committees, including an advisory group for the Phoenix Police Department. He was a founding member of Arizona Clean & Beautiful. He was a leader in the Phoenix Futures Forum, a multi-year program drawing on a wide range of interests to agree on a vision for the community and figure out the steps to achieve it.

Jack's vast number of commitments didn't sit well with everyone on the SRP board.

"A couple of fellows on the board told him 'By golly, don't you get on any more boards,'" former board member Cecil Miller told a reporter in 1990.[10] "And, by God, he did. Somehow or another, Jack always rolls with the punches."

Protecting a Ribbon of Life

The San Pedro River is a rarity, a free-flowing watercourse with no dams. Running north 143 miles from Mexico to the Gila River, it's a crucial stopping off point for millions of birds that migrate every year from wintering grounds in Mexico and South America to summer breeding areas in the northern United States and Canada.

But as the area around the San Pedro developed in the 1990s, so much water was being pumped from the ground that it threatened to suck the river dry. The demand for water was driven by the growing town of Sierra Vista and the Fort Huachuca Army base next door.

In 1998, Jack was appointed co-chairman of a tri-national advisory panel to recommend strategies for heading off an ecological disaster. The group focused on the U.S. stretch of the river, known as the Upper San Pedro. The project was one of the first environmental initiatives tied to the landmark 1994 North American Free Trade Agreement, signed by the U.S., Mexico, and Canada. It focused on birds not only for their importance as a resource but also as a measure of a region's ecological health.

Experts first analyzed the physical and biological conditions that migratory birds need along the Upper San Pedro. Then focus groups and workshops took public comment on the draft study.

Finally, the thirteen-member advisory panel reviewed the study and comments and drew up a report. Jack described the group as "very eclectic." The mix included the former premier of British Columbia, a member of the National Audubon Society, a retired Army brigadier general, and the director of the Arizona Department of Water Resources. Jack's co-chair was Fedro Carlos Guillén Rodríguez, then the chief of public participation at Mexico's Instituto Nacional de Ecología.

"It's made consensus a little more difficult," Jack conceded, "but it's a strong group."[1]

The advisory panel issued its report in December 1998 in Montreal, and it was published the next year as *A Ribbon of Life: An Agenda for Preserving Transboundary Migratory Bird Habitat on the Upper San Pedro River.*

The recommendations included taking irrigated fields out of cultivation, adopting a strategy to reduce groundwater pumping, updating local plans to manage growth, and boosting conservation. But the panel didn't recommend closing Fort Huachuca, despite the urging of some environmental groups, because the base had made such progress in reducing its water consumption, which was 26 percent lower than in 1991.

Achieving a consensus was a feat, but Jack worried that the sense of urgency was lost. He felt compelled to add his own postscript to the report, emphasizing the limited power of outsiders to save the San Pedro. While the panel worked hard to raise the issues and possible solutions, its report couldn't change the course of events.

"Many Panel members believe that the river will survive only if the local leaders have the courage and creativity to give protecting the River the same priority and energy as promoting growth," Jack wrote. "The local leaders must develop strategies that will accommodate future levels of growth that will not destroy the river and that can be 'sustained' with acceptable environmental consequences. There are no Arizona models to emulate, only sad lessons from failed opportunities. The Panel wishes the local stewards God speed and good luck."

Jack was right: The report didn't have the power to preserve the San Pedro's ribbon of life. While there are stepped-up efforts to protect it, the river's future still hangs in the balance. In 2014, a proposed Sierra Vista development with 7,000 homes got a green light from the state Department of Water Resources — until a Maricopa County Superior Court judge upheld the arguments of opponents and revoked the certificate of water adequacy that the department had granted. The case was appealed in 2015, just as plans were moving forward for a giant 28,000-home development in nearby Benson. While the developer claimed that wells for the project were too deep to affect the river, scientists disagreed.

What a Blast

Ash trays were everywhere and smoke-filled rooms were the norm when Jack began his career. He smoked a pack and a half a day, and he was rarely at a meeting at Salt River Project without a cigarette in hand. Then, on July 14, 1984, he quit cold turkey. "I've known for some time that smoking was bad for me," he said. "It was just a matter of developing the right mental attitude."

He succeeded with the help of SRP's Fresh Start counseling program. Not that it was easy. "Believe me," he said, "the first few days I thought there were devils sitting on my shoulders telling me, 'This is the worst mistake you've ever made.'"

Jack told his story in the October 18 issue of *Pulse*, SRP's employee newsletter, to help publicize that year's upcoming annual Great American Smokeout, when smokers are encouraged to give up puffing for just one day, a first step toward stopping for good. "What's important," he said, "is that the effects of smoking are reversible, and after you quit you can start to recover."

Jack had no idea that his lungs were already too damaged. He would have two decades in the clear. But lung cancer would come for him.

As a former smoker, Jack wasn't about to clamp down on others. In a Q&A in 1985, he said SRP had no plans to outlaw smoking in the office.[1] He hoped any disagreements over smoking could be worked out department-by-department based on "goodwill and common courtesies."

His reasoning would have surprised those who suspected Jack of being a closet liberal. "My own view is that we're over-regulated in our society," he said. "I watch with dismay as the Legislature tries to solve each and every social problem by some new law; and I concluded that you can't do it."

Jack was especially concerned about overregulation in the production of energy. Environmental concerns, in his view, were needlessly bogging down projects — the Palo Verde Nuclear Generating Station, where SRP was a major partner, was a prime example.

Jack's belief in looking for common ground and win-win negotiations paid off in labor relations. After the 1978 strike, SRP experienced far smoother relations with its union employees. A new publication, *Labor Relations Update*, alerted supervisors to potential points of friction by outlining grievances that had gone to arbitration. The kickoff announcement of the newsletter included a very Jack-like observation to set a non-confrontational tone: "As with most situations, in the labor area management wins a few, and the union wins a few."[2]

When the secretary-treasurer of the state AFL-CIO, Darwin Aycock, retired in 1987, Jack wrote a note of congratulations and received a warm response from Aycock that concluded: "You are a great man and I wish you and yours the best always."

Karen Smith met Jack when doing contract research for SRP while she was working on a PhD in history. Jack would drop by to chat from time to time. When she finished her dissertation, he heard she was hunting for a job — and he offered her one in strategic planning.

"Jack had had a vision for the strategic planning department which was years ahead of its time," Smith said. He brought together people from a wide range of disciplines, including an electrical engineer, a hydraulic engineer, some staff with a business and public affairs background, and Smith, a historian.[3]

"What fun that was," Smith said, "what a blast."

They looked five to ten years down the road, working on questions like whether SRP should join larger transmission networks. Around 1988, Jack asked Smith to examine issues of work and family — well before the topic had gained much traction. "He was always looking ahead," she said.

Jack also thought the time had come to offer benefits for those in "untraditional family environments." When an interviewer asked him if this meant medical insurance for gay couples, he said, "Again, it's only a matter of time. Ten years ago I probably did not know any for-sure gay people. Now I know a great number of gay people. That's just part of what's happening in Phoenix and we will have to adjust to that."[4]

SRP was a whirlwind of growth when Jack became general manager in 1976, and the pace seemed like it would never let up. "We were so busy in the '70s and into the early '80s, keeping up with the need to add to the resources to serve customers, we didn't have time to do anything else," said Richard Silverman, a member of the executive team who later became general manager.[5] "We were opening up power plants and holding hearings to get environmental approval."

The Coronado Generating Station was one of SRP's bets on future growth, and it turned out to be a loser for Jack. SRP decided to build the three-unit coal-fired plant near St. Johns in northeastern Arizona in the early 1970s, when demand for energy was skyrocketing.

The first two units went on line in 1979 and 1980. Even before they produced a single kilowatt, a soft economy and increased conservation had put a dent in growth projections. SRP found a deft way to handle the extra capacity: It sold 30 percent of the station to the Los Angeles Department of Power and Water on a short-term basis — which would be traded back for part of SRP's share in the Palo Verde nuclear power plant when it went on line.

But that was just a pause before demand took off again. Jack was running SRP with a sure hand, and the results were stellar. Net revenues hit a record $160 million in 1982-83. They reached new heights, $188 million, the next year.

Urban growth was fueling the numbers. With SRP's property-based voting system — upheld by the U.S. Supreme Court in 1981 — the vast majority of the board still came from a farming background. But the land within SRP's service territory was turning from fields into subdivisions. Jack had been urging the board for years to cultivate good relations with surrounding cities.

His argument was reinforced when the numbers came out for fiscal 1984-85. In a historic change, water deliveries to urban land eclipsed those for farm uses, 53 percent to 47 percent.

All those new homeowners needed power. SRP added a record 34,715 customers for electricity in 1985-86, and it was ramping up for what appeared to be another round of explosive growth. Construction began on Coronado 3, with a hurry-up goal of finishing within five years. SRP employees were also scouting sites for the next coal-fired station.

The work force was growing along with the customers. SRP needed more space and there was a potential spot right nearby. Legend City, an Old West theme park, had gone through rocky times since it was founded on the Tempe-Phoenix border in 1963. Owners came and went, and there were repeated closures. Still, many locals had a soft spot for attractions like the Lost Dutchman ride and Cochise's Stronghold. The park was sputtering in 1983 when SRP quietly bought it and several adjacent parcels.

"We undoubtedly have acquired more property than we will need in the short term," Jack said. "You don't just buy what you want, you buy what's available…" SRP was positioning itself for the next century. It would have room to expand its headquarters operations. And, Jack predicted, there would surely be development opportunities for land that was close to Sky Harbor International Airport.[6]

Legend City shut down in September, when the property's leases ended, and the site was razed. A local columnist had fun picturing the Utilityland that SRP might build there.[7] SRP wasn't going to build a theme park, but its plans were equally ambitious: Papago Park Center, a $300 million headquarters complex surrounded by private development. SRP's five-building complex would be the core of a 478-acre project that might include a research park, shopping, a hotel, businesses, and condos.

Planning for the first phase, SRP's information-services building, began in 1985. To start off on a high note, it was designed as a showcase, with strong horizontal lines and a soaring atrium. SRP staff took to calling it the Taj Mahal.

For Jack, it was more like a white elephant. In August 1986, he had to tell the board of directors that half the design work for the first phase, costing $400,000 to $500,000, would have to be thrown out.

"We wasted some money; there is just no other way of describing it," he said. The problem was that the work on Phase 1 had moved much faster than the master plan, and there were concerns that critical elements, such as heating and air conditioning, wouldn't mesh.[8]

At the time, Jack was juggling his work load, civic commitments, and a new level of responsibility as president of the Board of Regents. Some SRP board members were looking dubiously at how much time he devoted to other matters. "He spent a lot of time outside," said Bill Schrader, an SRP board member for more than a half-century. "Certain days he'd be coming in kind of late in the day."[9]

And yet as far as Schrader could see, "He was still staying on top of things."

Among the highest honors in Phoenix were the annual man and woman of the year awards, given by the Phoenix Advertising Club to recognize community service. Jack was a sure bet to get the award when he was nominated in 1988. The list of his organizations and activities took an extra page and a half besides the space available on the nomination form.

Many in the community would have been surprised at how many different groups had gotten a helping hand from the chief of a giant utility. As the nomination pointed out, "Jack's greatest achievement is his ability to do so much for so many people without seeking the visibility that could accompany such an incredible amount of service."[10]

Jack's speech at the awards banquet in February 1989 was touching, self-effacing, and inspiring. As always, he spread the credit, using the metaphor of himself as the face of a clock. Behind it was the mechanism that makes it all possible and too often goes unappreciated. "So tonight, I would like to acknowledge the support of the SRP team and to let you know that by honoring me, you honor the Salt River Project and my fellow employees.

"When Pat told our son, Scott, about the award, he said, 'Well, that makes me son of the year.' Indeed it does. Just as it makes our daughter, Suzanne, daughter of the year. It also makes my wonderful wife and constant supporter, Pat, wife of the year....So by honoring me, you also honor my family."

Rep. Mo Udall embracing Pat Pfister and Jack at one of the many events they attended together.

People would ask Jack why he was so active in the community. "There is no easy answer to that question," he said. He listed three basic reasons: continuing a family tradition of service, giving back to a state and community that had been good to the Pfister family, and a conviction that personal growth comes from helping others. "I measure my wealth differently than most," he said. "At the end of the year, when I take stock of my balance sheet, I look at the human qualities rather than the financial ones. If, after examining the assets, I have improved as a human being — then I have added to my personal net worth."

◇◇◇

Jack was at the peak of his career at SRP and at a high point in public service. Just as his leadership graph predicted, though, there was a cliff ahead. The stock market crashed in October 1987, rattling the national economy. By 1988, Arizona was in a severe slump, with no signs of recovery. Adding to the pressure, the power industry was going through an unprecedented spate of competition, consolidation, and uncertainty. SRP no longer had a comfortable monopoly on power, as other utilities expanded their territory. Natural gas and co-generation facilities, which produce both electricity and heat, were becoming tough competitors.

When SRP decided to push ahead with the construction of Coronado 3, with its $670 million price tag, no utility had extra electricity to sell on a long-term basis. Now the market had a glut of power.

SRP had sunk $110 million into building Coronado 3. It would cost another $140 million to close out pending contracts. But in February 1988, Jack recommended putting the project on ice until 2004.

"Continuing construction of Coronado Unit 3 would make us just one more utility with excess power in the Southwest," he said. "Sometimes it's better to buy than to own."[11]

It was hard to walk away from an investment of a quarter of a billion dollars, but the board voted to do it. The math was obvious. SRP could get long-term contracts from two other Arizona utilities for 6.1 cents a kilowatt-hour — 3 cents less than Coronado 3's power would have cost.

Ratepayers would pick up the tab for changing strategies. SRP cited the cost of Coronado 3 when it raised rates 5.6 percent four months earlier. Now the additional revenue would pay for mothballing the plant. SRP also called a halt to developing coal mines in New Mexico, at a cost of $2.6 million.

Months later, Jack had another expensive message for the board. The job of designing the first phase of the headquarters project was "out of control." SRP had asked for so many late changes in the information-services building that the architects were charging another $2.3 million on top of their $4.4 million original fee.

"We can use this as a learning experience," Jack said. "One lesson we learned is that until we know exactly what we want to do, we ought to do nothing."[12]

The lesson, however, wouldn't be used anytime soon. Plans for the other buildings had been put on hold in March, as SRP worked out its future size and direction.

The next year, 1989, SRP created a separate subsidiary to handle the private development at Papago Park Center. The proximity to the airport was never quite the spur that Jack imagined, although two light-rail stations eventually were built near the development. He had always known the project would take decades to unroll. A quarter century later, space was still available.

Soaring Away

Layoffs were coming at Salt River Project in the late 1980s. There was no way around it. Financial and competitive pressures were squeezing the power industry, and utilities around the country were shedding jobs. At SRP, the number of new customers, which had been running thirty thousand a year through the middle of the decade, had plummeted to fourteen thousand. Jack knew there had to be cutbacks. But where he expected to use a scalpel, he would be forced to take up an ax.

The SRP board approved a long-term business strategy in late 1987. The emphasis now would be on efficiency and lower costs. A process called SOAR — SRP's Organizational Assessment and Renewal — was launched to design a restructuring plan to meet those goals.

Jack saw SOAR as a preemptive move that would head off massive job cuts. In February 1989, he said preliminary estimates were to eliminate 250 jobs, which would be handled by voluntary severance packages, early retirement, transfers, and layoffs.[1]

In a company Q&A published March 2, he noted that other utilities had recently gone through painful layoffs. Arizona Public Service had chopped 840 positions. "These are the types of dramatic cuts we're trying to avoid. I believe we anticipated competitive challenges early enough so that we'll be able to manage the transition with minimal hardship to SRP employees…"

As for his own future, Jack said he expected to retire in three to five years.

SRP used the Cresap consulting firm to work with employees on an exhaustive analysis of staffing. Karen Smith was on one of the half-dozen teams assigned to look at various parts of the utility.

As they soon discovered, she said, "The structure had gotten out of control."[2] A supervisor might oversee just two employees. Offices had been set up for temporary needs, such as analyzing Cliff Dam, and were disbanded slowly, if at all. SRP had its own engineering division, which had nothing to do when projects were put on hold.

Even in the recent tough times, the work force had been growing by 4 percent a year. With a monopoly market and years of healthy revenues, "they'd never really had to make hard choices," Smith said.

SRP didn't have a few more employees than it needed — it had hundreds extra. "When we made that presentation to Jack and his senior staff, you could just see the color drain from their faces," she said. "He didn't think there would be that much that you would see as 'fat.'"

Up to one thousand jobs were fat, according to Cresap. Employees were staggered when the preliminary recommendations came out in mid-March 1989, just weeks after Jack's reassuring Q&A.[3] Shocked himself, Jack apologized for inadvertently misleading them. "During the past six months, I repeatedly said we did not plan any large-scale layoffs," he said in the *Pulse* employee newsletter. "That was true. We did not anticipate anywhere near the number which the study teams recommended."

Jack reassured the staff that there wouldn't be another round of SOAR if it were done correctly — another assurance that haunted him later.[4]

The union had no doubt that Jack was caught by surprise. "I believe he was not out to deceive us," said Dan McKinney, business manager of Local 266 of the International Brotherhood of Electrical Workers. "I don't think his credibility will be affected, from what I hear. He's been up front with us."[5]

Not everyone felt so receptive, however. Jack kept an anony-mous note addressed to him that said, "You are NOT treating SRP employees fairly…SRP employees believe that they have been reduced to being a number not a person. Your credibility is at an all time low."

The final cuts at SRP weren't as deep as Cresap suggested. The SRP board voted in May 1989 to eliminate 791 positions. Still, that would shrink its 5,800-person work force by over 13 percent. The impact was cushioned because the cuts included about 200 vacant positions, and 131 people had taken voluntary severance. The layoffs would be phased over a year.

As SRP pared back, Jack saw with painful clarity that the Office of International Affairs would have to go. When SRP's most basic operations were under the knife, it was becoming impossible to justify a department that didn't directly contribute to the core mission.

Jack broke the news to department head Ed Kirdar: His position and department would be eliminated, and he had the option of taking a retirement package. "My heart is with you," Jack told him, "and I enjoyed it."[6] But Kirdar's very visibility and success had become liabilities. His name kept coming up at executive management meetings, where everyone wondered: If we have to make cuts here, why are we helping Egypt?

Kirdar argued economics: The program was cost-effective. While SRP toured foreign visitors around its facilities for free, it made money from the fees when Kirdar gave presentations and training programs. But the problem wasn't just dollars and cents, Jack explained. SRP was tightening its scope and goals, as well as its finances.

"One of the most difficult decisions I have ever had to make was the one to eliminate OIA," Jack wrote in a note after Kirdar retired. But if he had spared a department that had been clearly identified as outside SRP's core mission, it would have been impossible to get employees to accept the other tough reductions of the SOAR process. "It just had to be," Jack wrote. "I know of your disappointment, but you should be very proud of what you were able to achieve."

The SRP brochure promoting the Office of International Affairs had a picture of a project canal on the front and a view of the planet on the back — SRP was now going to stick to the first image.

In public, Jack always spoke highly about Cresap, but in Richard Silverman's view, "I think he felt that the consultants had pulled the wool over his eyes, and that they knew all along what was going to happen; they were not candid with him."[7]

He felt betrayed. "There was not a dishonest bone in his body," said Silverman, an SRP executive under Jack who later became general manager. "When he told employees that he had been assured that attrition would take care of it, he believed that."

Jack never got over the experience. "He would tell stories of being at Sky Harbor, for example, getting his luggage from a carousel, and the wife of an employee who had been laid off screaming in public about how he had ruined their life," Silverman said.

The very next year, SRP was looking at layoffs again. Its position as the economical alternative to Arizona Public Service was in jeopardy as its competitor slashed costs. Unless SRP trimmed back, too, the price differential would disappear within six years. Now Jack conceded that SRP should have gone further with its cost cuts in 1989: "I think we did what we thought was the right level at the time. In retrospect, it probably was not enough."[8]

Jack was particularly alarmed as APS entered into a cooperative deal with PacifiCorp, which had been trying to buy its parent company. (SRP had made its own offer to buy part of APS territory and was rejected.) The details were complex, but the bottom line was increased competition to gain power customers.

Jack expected to stay through the SOAR restructuring process and wanted to give ample notice of his retirement, creating a transition period for the next general manager. What he didn't consider was how the early announcement could make him a lame duck — or spark a move to push him out the door ahead of time.

As the 1980s wrapped up, Jack mentioned retirement but never set a time frame for his plans. Board member Bill Schrader remembers hearing him talk about the leadership performance curve.[9] As he approached fifteen years, Jack told Schrader, "My curve is going down. I think it's time that I start thinking about getting out."

Jack revealed his target for retirement in an offhand way at a meeting of managers in January 1990. It was a Q&A session, and someone asked if he planned to run for governor. "No," he answered, "My wife wouldn't let me even if I want to. Seriously, I have no political ambitions.

"I do intend to stay at Salt River Project until sometime early in 1992. Then I plan to retire, teach school, and do some writing."[10]

The comment seemed innocuous. But it was considered a slap in the face by the two board members in the audience, who had no idea of Jack's plans. "John Lassen looked at me and said, 'Did you know he was going to talk about retirement?'" Schrader recalled. "And I said, 'No, I did not.'"

Lassen was furious that Jack had gone public with a retirement schedule without consulting him or anyone else on the board.

The next morning in the company cafeteria, Lassen didn't use subtleties when he saw Jack. "You know, Jack," Schrader remembers hearing him say, "I've been thinking about this. I think you need to leave as soon as possible. You'd better leave within a year."

Jack didn't say a word. "His face just fell," Schrader said. "It looked like the old saying, like he was gut shot."

Another blow came in September 1990. SRP posted its first loss since 1947. Revenues topped a billion, but the bottom line took a hit from prolonged outages at Palo Verde Nuclear Generating Station and increases in noncash expenses, such as depreciation. Jack called the loss "a one-time thing."[11] Board members, though, were losing patience with the financial challenges.

In October, Jack officially changed his retirement plans. He announced that he would leave on July 1, 1991. He hoped a successor would be chosen by the end of the year, so the incoming general manager would play an active role in the latest restructuring program.

The layoffs continued with round two of SOAR. The board slashed 414 jobs in December, although that included 150 already vacant, and shrank spending by almost $700 million over the next six years. The outlook was just too anemic. "The Arizona economy will remain relatively flat during 1991," Jack explained, "and it will come out of the economic downturn at a much slower rate than we were predicting just six months ago."[12]

The timing of the latest layoffs prompted a few employees to distribute "A Collection of Newly Revised Christmas Carols," with jibes at SOAR and Jack. They included "Silent Night, SOAR Night," "SOAR Ride" and "Pfister the Showman."

Now the scramble was on to replace Jack. Half a dozen executives were vying for the top job. The board, true to form, wasn't going to look outside. Decades of experience at SRP were a plus. In fact, the decision went to the candidate with the longest track record at the utility — and the one with the closest eye on the bottom line.

Carroll M. "Perk" Perkins, who came to SRP in 1956, was the top financial executive. He was just what the utility needed, observed one board member: "someone that knows how to track a dollar." He was smart, personable, and easy-going, but also blunt and not inclined to back down. In January 1991, the board voted unanimously to make Perkins general manager when Jack stepped down at the end of June.

Jack said he was pleased: "Perk is a very experienced and able utility executive and I will look forward to working with him on the transition."[13] In fact, he thought Perkins was the wrong person for the job — it was generally assumed that he wanted C.A. Howlett, who had extensive experience outside SRP, to succeed him.

Jack expected to act as a mentor for Perkins, just as his predecessor Rod McMullin had done for him. It never happened. "One of my major disappointments is that I offered to play that role for Carroll Perkins," Jack said his 1991 oral history, "but he was just not interested in it — not in the least."

Jack may not have realized how fundamentally his appointed successor disagreed with him. Perkins was candid, and a bit cutting, in an interview a few weeks after being picked for the top job.[14]

Looking at the most recent round of layoffs, Perkins said, "We went through it a year and a half ago. Mr. Pfister said we wouldn't have to do it again and then he had to eat his words."

He had no intention of following Jack's example of community involvement. He planned to spend more time with the company "taking care of core business issues."

"I'm more financially oriented than Mr. Pfister," he said. "I'm also a little more laid back — a little more people-oriented." Although Jack made a point of being accessible, Perkins considered himself more approachable. He painted Jack as an intellectual who intimidated many of those around him.

Since Perkins didn't need or want advice, there was no reason for Jack to stick around. He moved his departure up to April, maybe under board pressure.

At the start of March, Perkins put together his management team. He axed three top executives, including Howlett. He didn't want to tap Jack's knowledge and experience. He wanted to make a clear change of direction.

Government on a Diet

Gov. Fife Symington charged into office in 1991 with an ambitious plan to revamp state government, which he called bloated with "tax increases and runaway spending." The retiring head of Salt River Project was a logical choice to lead it. Jack Pfister would bring vision, people skills, and a pragmatic focus on results.

But it turned sour for Jack in less than a year. He got caught in a power struggle. Worse, he had growing suspicions that the process was rigged.

This was one problem he couldn't fix. All he could do was step away. And then, instead of presiding over committee meetings, he found himself called to talk with government investigators.

J. Fife Symington III had run for governor touting his strength as a seasoned developer. Although financial cracks were appearing in his empire, he won in a run-off against Phoenix Mayor Terry Goddard. In April, a month after taking the oath of office, he outlined Project SLIM, the acronym for State Long-Term Improved Management. The goal was to streamline the state's thirteen largest agencies and rein in their budgets. Education, although it consumed nearly three-fifths of the state budget, was left out because Symington had a separate task force tackling that area.

Symington chose Jack to chair the steering committee for SLIM. "He was really very good at holding his political leanings close to his chest," Symington recalled in an interview more than two decades later.[1] "So one never really knew for sure where he came down politically on some pivotal issues, which gave him a sort of a position of impartiality."

Project SLIM would begin by hiring a private consultant to develop the overall strategy. But it would also tap the knowledge of state employees and train them to carry out the changes.

The consultant had to be chosen through the state procurement process. Bids would be solicited through a document called a request for proposals. So Jack teamed up with Elliott Hibbs, the governor's fiscal adviser, to start drafting it. They developed a scope of work for the project after meeting with professionals who had done similar jobs in the past.

The timing was perfect for Jack. He was ready for a big project, although his calendar was rapidly filling with new responsibilities since leaving SRP. The belt-tightening and layoffs there had given him plenty of experience in organizational restructuring. He welcomed a plan that would soothe the pain of change by enlisting state employees to shape the process and become "missionaries" in carrying it out.

The structure for SLIM fleshed out quickly. Besides the fifteen-member steering committee that Jack led, Symington had appointed a twenty-five-person project team within state government. And the governor had created a new, paid executive position for SLIM, naming David St. John, an assistant director in the Department of Public Safety, to fill it.

The request for proposals to develop SLIM's strategy went out in summer 1991, and by late August, fourteen bids had come in, including ones from such heavy-hitting firms as Andersen Consulting, Coopers & Lybrand, and Price Waterhouse.

An evaluation committee had been set up to weigh the bids, and a state procurement officer, Robert Stephenson, explained to its members the strict rules of the bidding process. They had to sign affidavits stating that they had no conflict of interest and pledging to keep the evaluation process confidential and not to contact any bidders during it.

But Jack was worried about the makeup of the evaluation committee. It had too many state employees, he said, and too few members of the SLIM commission itself. Although Jack was on the committee, so was the governor's deputy chief of staff, George Leckie.

As a compromise some additional SLIM members were invited to attend meetings, but they couldn't vote.

Jack's uneasy feelings were, it turned out, right on target.

The evaluation committee met on September 3 to whittle the fourteen initial proposals down to four or five finalists. Under state procurement rules, the committee didn't have to judge bids solely by price. It could also consider which firm offered the best program for doing the job. In fact, committee members decided to toss out several of the very lowest bids over concerns that the firms didn't understand the breadth of work they were expected to do.

But dollars did count. Coopers & Lybrand had presented an acceptable plan and had worked for the state before, including consulting for the Arizona Board of Regents. But the firm wasn't on track to make the cut this time. Its bid of nearly $2 million was significantly higher than the others chosen for the final round.

The September 3 meeting was wrapping up when Leckie made a surprising, last-minute suggestion: Coopers & Lybrand should be included in the next round.

Leckie was someone to listen to. He was Symington's longtime confidant and his finance manager in the gubernatorial campaign. The men's friendship went back to 1975, when they both had children at the private Phoenix Country Day School.[2] Both also had ties to Coopers & Lybrand.

The firm, through its Arizona partner John Yeoman, was Symington's personal and business accountant. Leckie had close contact with him during the governor's race, since Yeoman had served as campaign treasurer. Leckie later claimed he had revealed the Coopers & Lybrand connections to the committee, but Jack said there was no such disclosure before a consultant was chosen.

Leckie's pitch for Coopers & Lybrand picked up some support, and the committee decided to keep the firm in the mix. Jack saw no reason to object. "I didn't think they would be competitive financially," he said in a later interview, "but I was not against them."[3]

The finalists had a chance to revise their bids and make their "best and final offers" to the State Procurement Office on September 9. Four of the proposals were unchanged or given relatively minor tune-ups. Two firms — Maximus and Booz, Allen and Hamilton — stuck with their initial bids of $1,075,000 and $1,676,000, respectively. Price Waterhouse sliced off $20,000 to reach $874,828, while Andersen Consulting cut $50,000, for a final bid of $1,445,000.

Coopers & Lybrand stood out. The firm had slashed its original bid by $441,200 — a 22 percent reduction. The new bid was $1,533,000. Coopers had gone from an outlier to a contender.

Procurement officer Stephenson was known for his strict, by-the-book approach to contracting. As a matter of course, he looked at the final bids before giving them to the evaluation committee. He was stunned. The new Coopers & Lybrand bid was jaw-dropping in the world of procurement. In twenty-eight years of experience, Stephenson had never seen such a huge reduction in a bid.

Jack, too, was taken aback when he saw Coopers & Lybrand's new number. "I was startled at the amount of decrease in their bid," he said. It was an "incredible coincidence," he thought, that with no knowledge of the other finalists' bids, the firm just happened to make a big enough reduction in its proposal to be competitive.

His own vote went for Andersen, based on its proposal as well as its lower cost. "I felt that either Coopers & Lybrand or Arthur Andersen could do the job," he said. "They approached it differently and I had a preference for the techniques that Arthur Andersen were going to use."

Coopers & Lybrand won, with six of the eleven votes cast. Jack didn't question the firm's competence, but he was sure it wouldn't have landed the contract without that big cut in its bid. Something was fishy, he thought, a feeling that several committee members quietly shared.[4] Jack talked with Hibbs about his doubts and the way Leckie had seemed to be pushing Coopers & Lybrand from the very start. Could he have passed on inside information?

Jack had no evidence to act on. "I suspected it," he said later, "but I did not know for sure."

Hibbs and another Symington staff member went to the governor with the concerns that Leckie may have contacted Coopers & Lybrand. The governor told them to ask Leckie directly about it, which Hibbs did. When Leckie denied any improprieties, Symington considered the case closed.[5]

Meanwhile, reporters had been taking a closer look at the winner of the SLIM contract. On October 16, *Arizona Republic* reporter Mary Jo Pitzl wrote an article pointing out that Coopers & Lybrand, through its accounting arm, had represented Symington on both business and personal matters. The firm's bid, she noted, was the second-highest of the five finalists. State officials, however, said there was no conflict of interest, since the bidding process was followed and Symington wasn't directly involved. They also noted that price was just one factor in awarding the consulting contract.

But Jack's suspicions that this was a fixed game were growing. In December 1991, Coopers & Lybrand filed a request to increase the scope of the project. The cost of the change was $400,000 — just about the amount the firm had cut from its original bid.

"They claimed that the scope of the work was different than what had been represented to them," Jack recalled. "They also claimed that the amount or the quality of the State employees that were going to be working on the project was different than they had anticipated and that they would then have to spend more effort. And I do not believe that they justified either one of those claims." Coopers & Lybrand's request was going through the governor's office, where, Jack was appalled to realize, Leckie was on the verge of approving it.

As 1992 began, Leckie had his own worries. He was now chief operating officer for the governor's office, and the finances there were a mess. There were flaps over payments to ex-aides and the expenses on a trade mission to Japan. Then came the revelation that Symington's office was overspending its budget and headed for a deficit.

Leckie lost his post in a staff shuffle. But he wasn't out the door. Symington named him director of Project SLIM, with an $85,000 annual paycheck. Of course, SLIM already had a manager earning more than $83,000 a year. Critics wondered why the project to streamline government operations had two chiefs, with combined annual salaries of $168,000.[6]

Jack asked to meet with the governor and Leckie to argue against the direction the project was going. He objected to putting Leckie in charge of SLIM because, in Jack's view, he was the wrong man for the job for a host of reasons. Jack didn't spell them out when he described the discussion a few years later. But he must certainly have questioned Leckie's financial management skills.

Then there was Cooper & Lybrand's request for the change in scope, with that fat increase in compensation.

Jack said he understood that Leckie was about to approve the change, and "I would not be, could not be party to that at all."

If Leckie remained in charge of SLIM, Jack told the governor, there was no choice. He would resign.

Jack had no clear evidence of impropriety at that time. "I was just uncomfortable with this change order for four hundred thousand which was near the amount they had originally reduced their bid."

Symington stuck with his decision, and Jack resigned on January 29. It was front-page news. One of Arizona's leading civic and business leaders had left the governor's premiere project.

Jack, as usual, was circumspect. Although his resignation came the day after Leckie's appointment to run SLIM, he declined to say whether that triggered his resignation. He would only say that since the project was diverging from its original concept, "I thought it would be better for the governor to select someone more comfortable with the current approach."[7]

At the time, Symington described the dispute as "a difference in philosophy" and an argument over control. Looking back, he recalled "a tug of war" in the SLIM committee, with Jack expecting to run it independently. But the governor's office had to have oversight on such a key initiative, Symington said, and "You can understand that I would want someone allied with me to head it."

Two days after Jack left, another member of the steering committee announced his resignation. Bill Jamieson, former director of the state Administration and Economic Security departments, said the process had become a "sham." He lost faith in SLIM, he said, when he began to suspect that Coopers & Lybrand had already set a target figure for cost cuts, before any study or discussion.[8]

The original implementation plan, as Jack understood it, had also been altered. Instead of letting state employees carry out the recommendations on their own, the grassroots approach that appealed to him, the consultant was going to be heavily involved in carrying out changes.

The whole experience was excruciating. "Jack valued his reputation very highly," Hibbs said, "and he was not going to do something or be a part of something that would cause people in any way to look at him differently."[9]

In Symington's view, "Project SLIM was very successful," bringing significant savings and the "total quality management" approach to state government. Skeptics saw the gains as little and short-lived. A report after SLIM's first calendar year claimed savings of $166 million, but SLIM's annual reports later put the figure at $29.7 million over two fiscal years.[10]

While the project counted job cuts of 1,300, seven years later, the government workforce had expanded by 6,500 positions and the state budget was $1.6 billion bigger, at $5.1 billion.

For its part, Coopers & Lybrand got expanded contracts, and compensation, but not as much as it sought. Then, in 1994, SLIM was back on the front pages of local newspapers, and the news supported Jack's suspicions, and worse.

The *Arizona Republic* reported in March that Leckie had phone contacts with Coopers & Lybrand accountant Yeoman during the bidding process, despite having signed a pledge not to contact any bidders.

The Maricopa County Attorney's Office led an investigation, at Symington's request, and found slipshod practices — such as allowing evaluation committee members to take documents home — but no evidence of criminal wrongdoing. Investigators said they couldn't prove the conversations between Leckie and Yeoman were illegal bid-rigging.

Symington was re-elected that year. By then, the state attorney general was looking into SLIM and had testimony from a former secretary that Coopers & Lybrand had inside information before submitting its final bid.[11]

Jack was called to be interviewed by the attorney general's office in April 1995. Three months later, the state reached a settlement with Coopers & Lybrand, which paid $725,000 while conceding only "at minimum a grave appearance of impropriety." Leckie also settled for $25,000 without admitting wrongdoing.

Now the feds were involved, too. Jack received two subpoenas to testify in the case. In March 1996, a federal grand jury indicted Leckie and Yeoman on criminal counts of fraud. Yeoman, who also faced charges of perjury and concealment, died a month later in a car crash. The next year, a jury acquitted Leckie.

Prosecutors were barred from introducing a sworn interview from Cooper & Lybrand's internal investigation, in which Yeoman said he had received confidential bidding information from Leckie. In a last-minute ruling, the judge held that the statement, which contradicted Yeoman's not-guilty plea, was inadmissible because it was made under duress (by cooperating, he could leave the firm without losing his benefits) and Yeoman's death precluded any cross-examination.[12]

Coopers & Lybrand itself ducked criminal charges by agreeing to cooperate with federal investigators still probing SLIM and the governor's finances. The settlement included a $2.275 million fine, community service and ethics training. The firm did acknowledge evidence of bid-rigging on SLIM, putting all the blame on Yeoman. Discrepancies in Symington's financial records, which Coopers & Lybrand audited, were also at issue: The firm pointed the finger at Yeoman and the governor.[13]

The deal was one more blow to Symington. His financial wheeling and dealing had been under FBI investigation for nearly five years.

On June 13, 1996, he was indicted on federal felony charges that included bank and wire fraud. Two weeks later, Jack and Pat Pfister found an appeal from the Gubernatorial Legal Expense Trust in their mailbox, asking them to contribute to the governor's defense. Jack's papers show no indication that he responded.

Symington was convicted the following year and resigned. (His conviction was overturned on appeal in 1999 over the dismissal of a juror. Federal prosecutors never retried the case: President Bill Clinton pardoned him just before leaving office in 2001 — Symington had saved the future president from drowning in their college days.)

For Arizona, it was one more governor leaving office in midstream. On September 5, 1997, Secretary of State Jane Dee Hull was sworn in as governor.

For Jack, it was another call to civic duty. Hull put together a transition team, and there was no question that he would be on it.

The Smart Moves

Phoenix was the national poster child for urban sprawl in the late 1990s. Bulldozers tore through cactus and old citrus groves as development devoured more than an acre an hour of desert and farmland. Roads were jammed, residents were routinely choked by a brown cloud of pollution, new schools were overcrowded the day they opened. The population of the metro area shot up 150 percent from 1970 to 1993, reaching 2.5 million, the third-highest growth rate in the nation. The population was expected to double to 5 million by 2025.

Jack had dealt with the side effects of growth his whole career. He played a role in expanding water supplies through the Central Arizona Project, meeting the rising demand for higher education, and keeping SRP on pace to handle climbing power use. Now Gov. Jane Hull was about to put the whole growth challenge in his lap.

Hull's official swearing-in ceremony was on September 8, 1997. The new governor intended the event to be low key, since she reached office through the disgrace of Fife Symington. But it turned into a celebration anyway. U.S. Supreme Court Justice Sandra Day O'Connor, who had grown up on an Arizona ranch, administered the oath of office, and more than a thousand people showed up for the ceremony on Capitol Plaza. Hull was flanked on the stage by legislative leaders, the attorney general, the wife of ailing former Sen. Barry Goldwater, and Jack.

Hull had met Jack some three decades earlier, not as a high-powered business leader but as a constituent in her legislative district. Hull was running for the Arizona Legislature and drumming up support in the neighborhood. She knocked on one door and ended up making her pitch to the Pfisters. Over the years, Jack became a trusted advisor, and she turned to him when she suddenly had the state's top job. "He was a great deal of help in my administration," Hull said. Jack could consider the broader perspective in a way that staff caught up in day-to-day grind couldn't. "He had absolute patience," she recalled. "He was continually mulling things over."[1]

Hull faced two immediate challenges: school funding and growth.

Arizona's property-tax system had created enormous disparities in money the school districts had available for capital expenses. While one elementary district in Phoenix couldn't afford to repair roof leaks, wealthy districts were building indoor swimming pools and even a domed stadium.

Dozens of poorer districts sued over the inequities in 1991, and in 1994 the Arizona Supreme Court ruled that the method of funding schools was unconstitutional. It ordered the Arizona Legislature to create a better system, but lawmakers repeatedly failed to agree on anything acceptable to the court.

When Hull took office, she immediately put a committee to work on finding a solution. The members were mostly staff from the Legislature and her office, including Jack. They met, in his preferred style, with a wide range of stakeholders from around the state.

"It became very apparent," Jack said, "that there was not an easy way to equalize the funding" — unless the state itself took over the responsibility of paying for school construction, maintenance, and other capital expenses.[2] Working with Hull, the committee came up with a plan called Students FIRST (Fair and Immediate Resources for Students Today).

Legislators were leery about putting a new financial burden on the state, but they approved the proposal in April 1998. Hull signed it — and the Arizona Supreme Court unanimously rejected it in June. The plan would still result in unequal funding. The exasperated justices gave the Legislature an August 15 deadline to come up with a better version. Otherwise, they would freeze all school funding.

Hull and her staff quickly revamped the plan. She called a special legislative session in July, and the tuned-up Students FIRST passed with a two-thirds majority, putting it into effect immediately after she signed it on July 9. The state would spend $374 million a year to build, equip, and maintain schools. The court issued a one-page ruling on July 20 closing the case.

Jack's name may not show up in the news stories, but Hull credits him with helping to get the measure passed. Certainly he had the persuasive power and the contacts to sway lawmakers. He was also deft at defusing panic and putting a crisis into perspective for young staff members. "He had a maturity," she said. "The bottom line was: The state's going to go on."

For a time, Jack worked out of the governor's office with the title of "special advisor." Maria Baier, Hull's advisor on growth issues, realized with a twinge of guilt that her own space there was larger than Jack's. Not that he cared or needed a lot of room. "You'd go into Jack's office," she said, "and it was clean all the time: no clutter, no stacks of papers. I'd say, 'Jack how can you do that?'"[3]

His answer explained one reason he was able to accomplish so much: "If I don't read it the day that I receive it, I throw it away. I'm not going to read it the next day. Tomorrow's not going to have any more time."

And Jack needed plenty of time for his other big issue in the Hull administration: growth. As Arizonans looked at ways to rein in urban sprawl, many saw Portland as a model. The Oregon city had adopted an "urban growth boundary," keeping development within a defined area and protecting the rural landscape outside it.

Environmentalists were leading a drive to create urban growth boundaries in Arizona, along with other new planning tools. They called their package of proposals the Citizens Growth Management Initiative and were gathering signatures to put it on the November 1998 ballot.

Jack saw the initiative as too extreme. But he knew that increasing numbers of the state's residents were fed up with the galloping pace of growth, the loss of open space, and the erosion in their quality of life. Jack helped the governor shape her own solution to the state's growing pains.

One wild card in any attempt to control growth was state trust land, which makes up 13 percent of Arizona. This is land that was granted to Arizona at statehood by the federal government as a way to raise money for schools and other public institutions. By law, it had to be managed for maximum income. And the biggest dollars came from development.

The dilemma was that trust land included areas that most Arizonans agreed should be off limits to development, such as the rugged desert of north Phoenix, the foothills of the fabled Superstition Mountains, and the grasslands of southeastern Arizona. But legally, the only way to protect them was to buy them at full-market value. And no one had the money, especially for land near Arizona's booming cities and towns.

Under Governor Symington, the Legislature had passed the Arizona Preserve Initiative, which created a way for trust land near urban areas to be bought or leased for conservation purposes. But without any money attached to the program, only a few slivers of land could be protected.

The Citizens Growth Management Initiative didn't deal with trust lands. Hull's plan would.

The governor announced her package of proposals — dubbed Growing Smarter — at a news conference in late March 1998. The plan included a requirement for cities and counties to adopt a general plan every ten years, with provisions for open space and growth corridors. The governor's package would provide state matching money to protect trust land.

Meanwhile, the pace of growth showed no signs of letting up. A staggering ninety houses a day were going up in the Phoenix metro area as the number of building permits hit record levels. The tiny Higley elementary school district southeast of Phoenix was a prime example of the impact of growth. The district had a single school for its 262 students in a twenty-four-square-mile area that had always been rural. Now thirteen planned housing developments were expected to add ten thousand students to the district.

Yet Growing Smarter had a rough trip through the Legislature. Lawmakers tried to add draconian protections for private property, and Jack had to use all his political juice to help Hull get legislators on board. The bill, which she signed in May, included provisions for a Growing Smarter Commission that would recommend more tools for handling growth.

But legislators still needed something to lure voters away from the Citizens Growth Management Initiative. So they put their own measure on the ballot: A plan for the state to set aside $20 million annually for eleven years in a matching fund to conserve state trust land.

Just a week after Hull approved Growing Smarter, the threat that had driven legislators into action suddenly disappeared. Environmentalists couldn't make the deadline to put their initiative on the upcoming November ballot. Now they were aiming for 2000.

So voters saw only one growth-related measure, funding for trust-land conservation. They said yes. And Hull moved ahead to set up the Growing Smarter Commission in November 1998. No one wondered who the chair would be: Jack, now a professor in the School of Public Affairs at Arizona State University.

The commission was composed of eight legislators, the state land commissioner, the state parks director, and five appointments by the governor. Besides Jack, Hull chose a land-use attorney, a rancher, a farmer who had been mayor of a small town in the Phoenix area, and Luther Propst, executive director of the nonprofit Sonoran Institute.

Propst debated before agreeing to participate. The Sonoran Institute has a broad mission to promote healthy landscapes, vibrant economies, and livable communities in the West. His would be the only environmental voice on the commission — on the other hand, otherwise there might be none.[4]

He quickly came to respect the chairman. "Jack understood public service," Propst said. "So many folks today don't. He was a kind person and a really good strategist. He would have made a great governor, knowing how to balance perspectives and keep people at the table."

Through the whole process, Jack spoke frequently with Hull. He worked closely with her staff, she recalled, not around them. He didn't feel any need to act the boss. "When you think of a major CEO," she said, "that doesn't happen very often."

Jack pledged that the commission would be "as inclusive as we can possibly make it." Hull had created a larger advisory group, with members ranging from environmentalists to ranchers, that would coordinate with the commission. And Jack put out the welcome mat to the public: "We will draw on anyone who has a constructive suggestion to make."[5]

Jack had one of his biggest challenges with the Growing Smarter Commission. He faced suspicions from both ends of the political spectrum. On the right were free-market conservatives with libertarian instincts, including some who saw "plan" as a four-letter, dirty word. Those on the left complained that Growing Smarter was too cozy with developers to make any meaningful proposals. Jack refuted both sides, saying the commission would seek "new ways for government to intervene without dictating" and still make recommendations that would "fundamentally change" the way the Phoenix metro area grew.[6]

The attacks on the commission weren't just rhetoric. Legislators proposed at least twenty bills that would severely limit or destroy the options for managing growth. Jack tried to head them off, warning that "Any legislation that would preempt the deliberation and consensus building now under way would be harmful."[7]

At the Growing Smarter Commission, Jack knew how to pace meetings. "You could just see Jack looking around the table to make sure everyone spoke their piece," Propst said. He would let the conversation continue long enough to reach agreement, if that were possible, and otherwise move along.

Breaking a problem down into smaller components was one of Jack's effective techniques. He divided the commission's work into eight subcommittees. Their proposals, presented in mid-May 1999, included giving counties the same land-planning authority as cities, adopting consistent local development fees, and allowing local governments to limit how far they would extend roads and other services. Virtually all of the ideas were bound to be controversial — especially a proposal to protect state trust land.

The commission got set to hit the road in June and July, touring the state to get feedback on its draft proposals. But first Jack held a workshop for the group to figure out the most effective way to reach out to the public.

Arlan Colton, who represented the state Land Department at the time, had never seen Jack in action. "I was blown away," he said.[8] Jack organized and ran the entire workshop himself — leading exercises in how to collect opinions, jotting ideas on Post-it notes, and using a flip-chart. He went over the finest details, from making people comfortable to staying on schedule.

"I couldn't even imagine that someone who had been the CEO of a huge organization could be this far down in the weeds in running a workshop and doing it so expertly," Colton said. Jack explained that his daughter, Suzanne, who did these sorts of exercises for a living, had given him expert advice.

Jack and Pat hosted the group at their house for a final wrap-up meeting, Colton said, a gathering infused with the glow of their warmth and ebullience.

And then they were off on what they dubbed, in a nod to the Beatles, "Governor Hull's Magical Mystery Growing Smarter Tour." Jack was along nearly the entire time as they hit every corner of the state, holding meetings in schools, libraries, and other public buildings. Pat frequently came along. While Jack was helping run the workshop, she would oversee a kids' corner so parents could participate without a child tugging for their attention.

Whether in rural or urban areas, Jack found, growth "clearly is a hot topic everywhere we go."[9] But that was the sole point of agreement. After meeting with almost twenty stakeholder groups, he said, "I don't think there's any unity on any topic."[10]

There was no unity in the commission itself, either. Columnist Robert Robb, wrote jokingly in the *Arizona Republic*, "If you want to see the best juggling act in town, don't bother with the circus. Go see Chairman Jack Pfister (as he tries) to get a consensus out of his Growing Smarter Commission."[11]

Yet when the commission met in August 1999 to draw up its final recommendations, Jack managed to cobble together agreement on more than twenty smart-growth recommendations. They covered the spectrum of concerns, from creating incentives for infill and requiring better local planning to mandating written consent from property owners before rezoning property and paying ranchers for good conservation practices.

The one intractable issue was the proposed Stewardship Trust, a mechanism to protect some state trust land without having to buy it. Commission members couldn't agree on how it would work, how much land should be set aside, and who would decide which pieces to preserve. Jack solved the problem by ducking it. The commission would endorse the idea, but leave the details to a future task force.

Anything more precise would have jeopardized the final agreement on the Growing Smarter list of recommendations. "I would feel utter despair if we broke down over a few words," he explained. "We will have done Arizona a grave disservice."[12]

Now the Growing Smarter package went to the Legislature, where the reception from conservative Republicans was decidedly cool. Once again Jack had to play defense, arguing that here was a plan representing the middle ground. "We spent a lot of work trying to balance all the special-interest positions," he said, "and I hope the legislature will seriously consider it."[13]

Jack was also quietly working on the big picture. He still hoped to find a growth-management plan that the business community, legislators, and the Sierra Club and its allies would all support. His optimism came from his conviction that even on the most controversial of issues, the differing sides usually have some overlapping goals.

Sandy Bahr attended countless hours of meetings of the Growing Smarter Commission to represent the Sierra Club's perspective. In her view, Jack failed to see how much the battle over growth differed from the water issues where he'd been able to find consensus.[14] The governor back then, Bruce Babbitt, had the trust of environmentalists, a track record of supporting conservation, and the clout to get a consensus deal through the Legislature intact. Hull didn't.

In Jack's version of compromise, Bahr said, each side gave up a little. For environmentalists, however, an acceptable compromise had to move the issue forward incrementally and leave room for future progress. It couldn't set the status quo in legal cement.

Jack's hopes for a grand bargain were dashed. And the Growing Smarter Commission's plan failed to emerge from the Legislature intact.

Lawmakers did, however, approve a package of measures giving local government new powers and duties to manage growth. Dubbed "Growing Smarter Plus," it reflected some but not all of the commission's recommendations. Communities would now be required to adopt general plans every decade with specific points, such as strategies for efficient transportation and well-timed expansion of sewers, streets, and other infrastructure. Developers would have to pay their fair share of the cost of new services for their projects. Hull called it a "historic commitment to land planning."[15]

There was also a ballot measure, Proposition 100, to protect up to 3 percent of trust land, some outright and the rest through a system that included approval by a two-thirds vote of the Legislature.

Propst, the environmental voice on the Growing Smarter Commission, was disgusted. "What came out of the Legislature was not at all what went in," he said. "The process was dishonest." The governor and Jack, however, felt they had the best deal they could get. They leaned on Propst to endorse it, but he refused. "Jack was not happy with me," he recalled, "but he was a true gentleman every step of the way."

The business community, led by homebuilders, poured money into defeating the Citizens Growth Management Initiative and, as an afterthought, promoting Proposition 100. They outspent the other side by a margin of eight to one.

Faced with two apparently competing ballot measures — in fact, they addressed completely different subjects — voters did what they usually do. They rejected both.

So did Arizona end up growing smarter?

For all his criticism of the ultimate result, particularly regarding trust land, Propst thought Growing Smarter brought some improvements in planning. Unfortunately, he said, the Legislature continued trying to strip local governments of their ability to plan.

Lawmakers could have taken a lesson from Jack, according to Propst. "I remember thinking that if the Legislature had one-tenth the integrity and intellectual capacity of Jack, Arizona wouldn't be the laughing-stock of the nation."

Baier, Governor Hull's point person on growth at the time, said Growing Smarter has been particularly effective with its requirement for public approval of general plans and major amendments to them. "You can talk to anyone who does real estate development," she said, "and they will say that scrutiny has made a big difference."

While Arizona failed to adopt an overall plan to protect trust land — voters turned down a second attempt after they rejected Proposition 100 — the matching funds from Growing Smarter have had a big impact. With the money, Phoenix has set aside swaths of desert in the north, and Scottsdale has saved large parts of the dramatic McDowell Mountains.

"Jack was a seasoned veteran of public policy and understood that great things happen incrementally," Baier said. "I don't think Jack would sit here and say he was disappointed. He was a pretty positive guy. He had been through too many wars to think you're going to win all the battles."

The Most Satisfying Job

Jack had all the instincts of a great teacher. He had a passion for learning, he believed in the power of knowledge, he enjoyed working with students, and he had vast reserves of patience.

Whenever he looked at his future after Salt River Project, he had always talked of teaching, maybe at Arizona State University.

Now, faster than Jack originally planned, that retirement was coming. So in 1991, ASU President Lattie Coor approached the director of the School of Public Affairs, Joe Cayer, to see if there was any interest in having Jack join the faculty.[1]

Of course, everyone knew the men were friends and that Jack had helped recruit Coor to ASU. A newspaper columnist slammed the connection, and suggested it was a cozy deal of "you rub my back, I'll rub yours."[2] But he ignored Jack's qualifications. Jack could bring a depth of experience in law, business, public policy, education, and civic service that would be hard to match anywhere. Cayer and his colleagues were enthusiastic, and Jack was offered a position at ASU as a "distinguished research fellow." Besides teaching, Jack would work on projects at ASU's Morrison Institute for Public Policy.

Coor also tapped him to join a planning team that was figuring out the logistics of a university that would have several campuses — besides the core site in Tempe, there was already ASU West — and instructional centers.

Cayer was a bit thrown at the prospect of supervising one of the state's most influential figures. But that changed the minute he met Jack. "He made me feel totally comfortable." Cayer said. "He was just one of the most humble and nicest persons you can imagine."

He taught at least one course a semester while he was at ASU. The students were getting master's or PhD degrees in public administration, and Jack covered a wide range of topics, from running nonprofits to municipal and state fiscal management.

Cayer had to read all the student evaluations of faculty. Jack invariably got high marks.

"The students just loved him," Cayer said. "They found him very demanding, but because of his personality and the way that he dealt with them, they didn't mind."

They appreciated Jack's real-world experience and success. They liked the way he didn't simply lecture, but engaged the students. He also expected them to go outside the classroom and learn, and then share their insights.

Whatever class you took from Jack, you were bound to come away with one lesson: Respect.

One of his key messages was that managers need to respect the people they oversee and solicit their input. It shouldn't be a one-way process of command and control. "He also really was very strong on the idea that you respect one another in work situations," Cayer said.

He didn't have any sympathy for slackers. "Work hard to be happy," he would tell students.

For many, he was one of the professors they would remember forever as someone who influenced their lives. A student who took a night class from Jack on ethics in nonprofit organizations never forgot the discussions on leadership. "They have shaped much of my leadership principles and practices today," he wrote in an online condolence note after Jack's death.[3]

Whatever idea a student came up with, however far-fetched, Jack always reacted with interest and never had a word of criticism, said Kim VanPelt, who took a course from Jack around 1999–2000. "I think he saw teaching as a way of having a conversation, not so much instructing as learning from one another," she said.[4]

Jack's classes were particularly valuable because he explored practical steps for dealing with real-world challenges. One time, students were presenting strategies for the case of a large nonprofit that needed to reduce expenses by 5 percent. VanPelt proposed letting departments of the imaginary nonprofit decide how to trim back, with eliminating jobs as the baseline alternative. "I remember him almost wincing at the idea of staffing cuts," she said. With the layoffs at SRP still a painful memory, Jack urged the students to avoid cutting staff whenever possible.

But he wasn't telling students to be overcautious. His final advice to students stuck with VanPelt as a rallying cry when opportunities came up: "Be bold."

On a purely practical level, Jack knew how students struggled with the financial burden of going to the university. He and Pat created a legacy of support for students with the Pfister Family Scholarship, a fund they established in 1997 to provide aid to master's students in the School of Public Affairs.

While the classroom was his love, Jack spent much of his time helping Coor refine the vision of a multi-campus university. It was Jack who coined a phrase for it: "One university geographically distributed."

Jack retired from the School of Public Affairs in December 2000.[5] He'd spent nine years on the faculty. "Of all the jobs I have done, teaching has been the most satisfying," he said. "The student interaction is the thing I am going to miss the most."

At long last, Jack thought, he would be able to pursue a writing career. "This is something I have always wanted to do," he said. He collected books about Arizona history as a hobby. Now he was eager to write some of his own.

His retirement lasted barely a year and a half — not that he really had much free time, anyway. Jack was still heading the board of ASU Research Park, serving on various other boards, working on the follow-up to Growing Smarter, co-chairing the governor's water management commission, and advising state and local leaders.

Whatever time he squeezed in for writing, though, was about to disappear.

In 2001, ASU President Coor gave the traditional one-year's notice that he would retire the following June. But an unexpected problem came up. The vice president for Institutional Advancement announced he was leaving to take a similar job elsewhere. Behind the opaque title was a vital position that oversaw such key areas as fund raising, lobbying, and public relations. Coor knew that his successor would want to fill the post, but he couldn't afford to leave it vacant for an entire year.

Who could parachute in, hold everything together, and then willingly walk out the door a year later? Jack, of course.

"I always worried that I took advantage of Jack," Coor said a bit ruefully. Not that Jack even considered declining. "He didn't hesitate."[6]

For university staff, Jack brought some calm to an unsettling transition period. "He came in and was a great leader and put everybody at ease," said Nancy Neff, assistant vice president of public affairs at the time.[7]

Short as his time would be, Jack made it clear that he wasn't simply babysitting. For instance, Neff said, he had a list of employee salaries and somehow noticed that her pay was out of line with comparable positions. She herself had no complaints about her pay and no idea that there were any discrepancies. Jack went to the budget staff and had Neff's salary raised. "I was flabbergasted," she said. Despite all the new areas for his attention, he was still concerned whether someone's pay was fair and equitable.

"That showed me what an ethical person he was," she said. "It was one of the many little things he did to make people feel good and respected." At holiday time, Jack bolstered morale with a thank-you buffet, reaching into his own pocket to pay for it.

He had a sympathetic ear when Neff was wrestling with how to handle a teenage son who didn't really care for school. "You're doing your best," Jack assured her. "Kids have a way of working things out. You just have to be there." He shared similar situations he had faced. "That was the thing that always struck me about Jack," she said, "how he could command a room and make decisions and be that person who was a CEO type person and yet you could just have an average conversation with him about his kids."

When Coor left ASU at the end of June 2002, so did Jack. It was his third retirement, he joked. Finally, he hoped to write.

Family research had uncovered a step-great-grandfather who came to Arizona in 1882. He lived in Apache County in eastern Arizona, and as Jack looked into the territorial era there, he uncovered fascinating stories about land grabs and an organized campaign to drive away Mormon settlers. He'd gotten started on the writing and planned to call the history, "Run by a Ring." In addition, he was teaming up with an ASU professor, Brent Brown, to work on a monograph about Burton Barr, the powerful Republican majority leader in the Arizona House of Representatives, who had died in 1997. And he had plans to do a comprehensive history of Arizona from statehood on.

"I've done lots of research," Jack said. "I've done a little writing and I've concluded that writing is a full-time job, so I'm going to make it my next full-time job."[8]

Once again, Jack didn't have time on his hands for long, if at all.

Coor had been batting around a concept for years — an organization that would take good ideas out into the real world, an area where he thought universities fell short. He had come up with a name, the Center for the Future of Arizona. It would be a "do tank" rather than a "think tank."

Through Harmony Alliance, Jack had learned how to turn a gauzy notion into concrete reality. He understood the steps to setting up a nonprofit. With Jack's help, Coor founded the center in March 2002.

This time Jack didn't want a high-profile role, so Coor didn't ask him to join the board. Instead, he was almost a staff member, coming in at least one day a week and going to internal meetings.

"He was a central figure in our first project, Beat the Odds," Coor said. Arizona's high dropout rate was a drag on its future that was only going to get worse as even the most low-level jobs increasingly required the educational background to deal with technology and sophisticated equipment. The center decided to focus on strategies to increase graduation rates for Latinos, who would soon make up a majority of school kids.

The approach was vintage Jack: find high-performing schools, analyze the reasons for their success, and suggest ways to use those tools elsewhere. ASU's Morrison Institute co-sponsored the project. Rather than simply track down best practices, the project's research team followed the guidance of business guru Jim Collins, author of *Good to Great,* looking at matched pairs of schools, similar in every aspect except their level of performance, to see why one achieved so much more than the other.

The results were published in a 2006 report, "Why Some Schools with Latino Children Beat the Odds…and Others Don't." It identified six keys to success that were summed up as: a strong and steady principal; collaborative solutions; a clear bottom line; ongoing assessment; a "build-to-suit" focus on individual student performance; and a commitment to stick with the program.[9]

The next year, the center established the Beat the Odds Institute and a school partners program. When it decided to follow up with a how-to handbook, Jack edited the chapters, along with Heather A. Okvat, a graduate research associate there. He even co-authored a chapter with Okvat, discussing how to use a self-evaluation survey in the handbook.

The handbook wasn't published until 2010, the year after Jack's death. It's dedicated to his memory.

Right up to the end Jack maintained a regular presence at the Center for Arizona's Future. "Jack sat at our policy table the week before he died," Coor said. "Jack wasn't capable of decelerating."

Research in the Park

More and more, Jack was becoming Arizona's executive rescue leader. If there was a problem, he was the go-to person to help devise a solution. In 1990, Arizona State University President Lattie Coor called him with one giant save job: ASU Research Park, whose finances were wobbling badly.

University research parks had been all the rage in the 1980s. And Arizona State University had just the place to put one. It had closed down a 320-acre experimental farm nine miles east of the Tempe campus at Warner and Price roads. In 1983, the Board of Regents voted to approve ASU's proposal to turn the empty fields into a research park. (Jack abstained over a possible conflict of interest if Salt River Project should decide to do a similar project.)

Stanford University was the Eldorado every university hoped to copy. With Hewlett-Packard and other high-tech companies as tenants, Stanford's research park provided a steady stream of revenue for its endowment fund.

ASU Research Park broke ground in 1984. The concept had a mix of practicality and vision that was right up Jack's alley. The goal was to attract commercial tenants engaged in research and development that meshed with university programs, thus promoting fruitful and profitable collaboration. As tenants leased land to build facilities, the university would reap extra income. And as companies expanded, they would create jobs and spur growth.

But ASU was moving into a tougher, more competitive market than anyone expected. In the gold-rush fever of the 1980s, the number of research parks in the U.S. more than doubled. Then, in the late 1980s, the economy slumped, and so did the prospects for adding tenants.

It turned out that everyone had missed a crucial point about Stanford: Its research park was founded in 1951. Success came over decades, not overnight.

By the time Coor called Jack in 1990, ASU Research Park had gone through two directors in five years. It wasn't even remotely on track for the glowing early estimates of being fully occupied by 1992, with five thousand employees. At the end of 1989, less than 10 percent of the land had been leased, and fewer than five hundred people were working there.

Jack was about to finish his term on the Board of Regents, and he wanted to keep an inside link for the research park. So he asked fellow regent Andrew Hurwitz to join its board. The project didn't particularly interest Hurwitz at the time, he said, "but I think I'd reached the point where I never said 'no' to Jack." He stayed on the board for the next thirteen years.[1]

Jack proceeded as he always did. He made a plan. The first step was to hire really good real estate professionals and get their advice about how the park could go forward, given its legal constraints.

One big constraint was that the research park could only lease land, not sell it. Since the leases were for ninety-nine years, there was no practical difference between leasing and buying. But bankers, developers, and potential tenants were skittish about the unfamiliar setup, and Jack had to sell them on the value of the leases. Jack's strategy was to target local companies first, where an expansion into the research park would be an easy move. He played up the potential for them to make new connections with the university and faculty research.

Experienced companies, Sunbelt Holdings Management Inc. and PCI Associates Ltd., were hired to handle the park's day-to-day management and marketing operations. But Jack remained hands-on for the next fifteen years. He started off with the role of executive director and became president of the board of directors in 1993.

"Jack devoted an extraordinary amount of time to that job," Hurwitz said. "It was pretty clear that there wasn't anybody else who could do that job to the level that Jack would."

With Jack in charge, board meetings followed the pattern he preferred. First and foremost, Hurwitz said, there was "incredible solicitousness toward everyone's views." Secondly, Jack decided that the meetings should include some fun. So he would invite scientists from ASU, tenants, and prospective tenants to make presentations about their latest projects. Hurwitz remembered some researchers who came in and explained that they could make a telephone smaller than a paper clip. "What they were having trouble with," he said, "is they couldn't put a keypad on it."

The purpose was more serious than show-and-tell, however. "I think he wanted us to develop a vision of what we wanted to happen there," Hurwitz said. "And you couldn't really do that by just talking about what kind of lease you were going to enter into." He wanted board members to know what ASU faculty were doing and what was at the cutting edge, so they could brainstorm corporate synergies.

"As is true with a lot of things, Jack helped turn that thing around," Hurwitz said. Besides his skills, the park benefited from a pickup in the economy and the commercial real estate market.

It could still be a slog. One commercial real estate broker brought in thirteen possible tenants for land or space over a six-month period in 1992. None of them worked out. "Some of those were very active discussions," Jack commented. "But we wound up the bridesmaid instead of the bride." [2]

But two big catches came that year when Motorola chose ASU Research Park for both its western training site and the headquarters of its flat-panel display division.

Soon, Hurwitz could go to regents' meetings and report that the research park was no longer going to be a drain on the university's resources. Now the question was whether it could actually generate some dollars for ASU. "But we solved the problem we initially were so concerned about," Hurwitz said, "which is that the university was losing money off this each year, money that could be used in better ways elsewhere."

By the end of 1995, a business columnist could write, "The ASU Research Park was a dream that almost became a nightmare but which now seems to be fulfilling its vision."[3]

Leading the park would continue to be a roller coaster ride, however. Avnet Inc.'s computer-marketing group opened a new headquarters in the park in 2000. But the next year, Motorola decided to pull out and sell its building there.

The park reached the twenty-year mark in 2004, and it got a big birthday present. ASU landed a five-year $43.7 million grant for the U.S. Army's Flexible Display Center, which would develop low-power computer displays for use in the field. The program would be housed at the research park in the old Motorola building, which the university itself bought earlier that year.

The Army project was a big sign that the park had achieved financial stability. It was gradually filling out with research-oriented companies and corporate headquarters and had reached two-thirds capacity.

The university was planning to put some of its most advanced labs in the Army facility. This was the sort of collaboration that the research park was intended to foster.

"It was anticipated when it opened that there would be an ASU presence at the park, but it never really happened," Jack said. Now the dream was finally becoming reality.[4] Jack could see that the rescue mission had been accomplished and then some. He retired as president of the board of directors in 2005.

When Coor made his 911 call to Jack, ASU Research Park was in rough shape, its long-term survival in doubt. With Jack's help and guidance, it was on track to flourish, weathering the virtual collapse of the Phoenix real estate market after the 2008 financial crisis. When the research park celebrated its thirtieth anniversary in 2014, it issued a press release proudly announcing the land was 89 percent leased, and the park was home to forty-nine companies employing more than forty-five hundred people.

The Torch of Public Service

Paul Johnson was young, unknown, politically inexperienced, and naively ambitious when he decided to run for Phoenix City Council in 1983. He had no real shot at upsetting the savvy incumbent. And he made a string of errors when he spoke at a candidate forum.[1]

Jack was in the audience, and he must have winced at the mistakes. But he saw a spark of promise in the twenty-three-year-old. At the end of the gathering, he came over to Johnson's table and introduced himself.

"Let me tell you a little bit about some of the facts you had, some of the things that weren't exactly correct," Jack said. In a deft, friendly manner, he gave Johnson extra information and set him straight on a few points.

Then Jack invited Johnson to meet with him at Salt River Project and take a tour to see how it worked.

"It was really kind of amazing," Johnson said. He was a small contractor, with no connections to the establishment, the business community or a political party. Jack had to know that this kid had no shot at winning. "But he didn't care," Johnson said. "He knew that I was charged up about politics. He knew that it was something I wanted to do, so he helped feed it."

Johnson took up the invitation to visit SRP, and Jack not only took him around but also talked about some of the hot issues of the day. Then he urged Johnson to apply for Valley Leadership, a program to develop the talents of existing and emerging leaders.

Looking back, Johnson marvels that Jack carved out so much time for him. It was a civic investment. "I think he saw the advantage was that more people ought to be involved, more people ought to take part," Johnson said.

Johnson lost his first race, but he took Jack's advice and sharpened his abilities in Valley Leadership. When he ran again in 1985, he went to meet with Jack. By then, he'd already managed to visit a number of executives of major companies. When he walked into Jack's office, he was struck by how much more modest it was. He knew Jack didn't drive a fancy car. His glasses were heavy and practical, not stylish. "He wasn't splashy at all," Johnson said. "He just didn't look like a big, powerful CEO."

Johnson couldn't help bringing it up. Was it because of SRP's status as a public utility that Jack's office wasn't as nice as those of executives elsewhere? "One of the things you need to understand," Jack said, "is that all of those people are on opium."

"You mean like heroin?" Johnson asked.

No, Jack explained. He was talking about OPM, shorthand for "other people's money." When you use other people's money to enjoy a flashy style, remember that you'll be the target when things go sour. If Johnson won a city council seat (and he did, despite being an underdog), Jack cautioned, "Just remember, it's always other people's money. It's not yours."

Although Johnson was then a Democrat, Jack showed him the value of following a nonpartisan path. Jack, a lifelong registered Republican, advised him not to treat people by their party labels but as individuals. Eventually, Johnson officially became an independent — as Jack, in the end, had really done unofficially.

Johnson was appointed mayor of Phoenix, the nation's tenth-largest city at the time, when then-mayor Terry Goddard stepped down to run for governor in 1990. At thirty, he was the youngest mayor of a major U.S. city and was elected in his own right in 1991. Jack continued to be a mentor on the many issues Johnson faced.

One was a proposed land exchange. The federal government was trading its historic Phoenix Indian School site in midtown Phoenix for Florida wetlands owned by the Barron Collier Companies. The city of Phoenix, however, hoped to create a park that would include the old boarding school, where generations of Native Americans had gone through a controversial system of off-reservation schooling and assimilation. So the city offered to swap downtown land with Collier.

Johnson needed to persuade federal officials, Collier, and Phoenicians that the swap was worthwhile. Jack advised him to strengthen his position by getting the business community on his side. "As a Democrat, it wasn't natural for me," Johnson said. Jack taught him to enlist business leaders as allies in accomplishing his goals.

In the end, a three-way deal was arranged in which each side came out with the key things it wanted. The Interior Secretary and Collier signed off on their part of it at the end of 1992.

Johnson started his political career as an outsider who eyed big business with suspicion. His views were more confrontational and divisive. In Jack, he saw a different model, someone who looked for common ground and avoided bombast. Johnson ended up adopting that more open approach. "I've just found that style makes me happier," he said.

He also became one of many who picked up the torch from Jack. Now he makes a point of meeting with young people interested in politics, even if he doesn't think they have a chance.

What motivated Jack? "I think that in public policy, Jack was driven by the desire to get things done," Johnson said. "In his nature, he cared about people." He was so effective so often because he put the two traits together.

Rent-A-Dad

Jack didn't have an official title in the latest phase of his life. So journalists gave him one: "civic activist." It was exactly right. For all the times he proclaimed that he would devote himself full-time to writing, he never managed to do it. He cared about too many issues. His talents were needed in too many places.

"If he had a fault," said former Phoenix Mayor Terry Goddard, "it was that he didn't say no. If you asked him for an appointment, he accommodated you, no matter how trivial the matter. He had a huge sense of responsibility in his post-Salt River days. I don't think he went so far as to think he was the conscience of the community, but in fact he was. He was the person you went to if you wanted a conscientious reading of what should be done. If you didn't care about that, you stayed away from Jack."[1]

So many people called on Jack for advice, so many benefited from his mentoring, that family and friends affectionately referred to him as "rent-a-dad."

Phoenix voters elected a new mayor in September 2003. Phil Gordon trounced his only opponent, winning 72 percent of the vote. With a majority win, he didn't have to face a runoff election in November. He did have to face an unusually long transition period until he took office in January.

He'd gotten to know Jack while on the Phoenix City Council and later sought his advice during the campaign. So when he needed to pick a leader for the transition team, Gordon said, "I never even considered anybody else."[2]

The mayor is designed to be a nonpartisan post in Phoenix, but party affiliations are no secret. Gordon was a Democrat, while

Jack was a registered Republican. But it didn't matter. He knew Jack wouldn't get hung up on ideology but would look at the practical next steps in governing the city.

He also knew Jack's unshakable integrity. "With Jack," he said, "it was not whether he agreed with you, but whether you were standing on principle" and not taking a position for the political payoff.

Jack helped line up five members for the transition team. He made sure that none had any business ties to the city or other conflicts of interest. He didn't say so, but Gordon knew that it would be a mistake to suggest anyone simply on the basis of being a friend or political ally. Jack was "like the father that you never wanted to disappoint," Gordon said.

Gordon's platform had the standard planks of education, public safety, and jobs. He also had a grand plan that was so far only sketched on a napkin: a downtown university campus. Jack would help make it happen.

Arizona State University President Michael Crow, who succeeded Lattie Coor, had visions of creating "the new American university," an institution that would become embedded in the community, learning from the local experience, and contributing to it. At the same time, he had a rapidly growing student body that needed more space.

Crow and Gordon met for breakfast during the campaign. The ASU president wanted to expand into an urban setting. The mayor wanted to bring residents, jobs, energy, and the cachet of a university to a central city that still needed a jump-start. At the end of the meal, they had outlined a downtown campus on the back of a napkin.

Jack helped Gordon sell the project in the toughest of times, when Phoenix was suffering through one of its worst economic slumps. "He taught me to speak to the business leadership in the language they would understand," Gordon said, making the case in dollars-and-cents terms.

Jack met with the downtown movers-and-shakers. He came along with the mayor to meetings to generate grassroots support, because the project would require Phoenix to provide land, buildings, and infrastructure.

"He was detail focused," said Gordon, who was notorious for wandering off point. "That is not one of my strengths. He made sure that we stayed on track."

In March 2006, Phoenix voters approved a bond measure for typical city services, such as parks and libraries — plus $223 million to develop ASU's downtown campus. It was, as far as anyone knew, the first time that a city government had financed a state university's expansion. That fall, more than three thousand full-time and six thousand part-time students were taking classes downtown. In less than a decade, the campus was home to journalism, nursing, health care, letters and sciences, and public programs. The law school was relocating there.

Jack's role, as in so many issues, was mostly behind the scenes. Nothing marks his contribution to the success of ASU downtown. "That's one of the names that never got the credit it should have on that project, which is one of the most important things that Phoenix has done in decades," Gordon said.

The list of Jack's low-profile service included an association with the Nature Conservancy that went back to the mid-1980s. His contributions there are a microcosm of what he brought to so many organizations.

The group wasn't very well known then, and around 1984, state director Dan Campbell decided to introduce himself to the movers-and-shakers of the Phoenix 40.[3] He was particularly eager to tell them about the new natural heritage database, which included a record of areas vital to wildlife.

Jack took up Campbell's invitation to check out the data and map showing critical habitat areas. "The most interesting thing to me," Jack said, "isn't where they are, but where they aren't." He immediately saw how Salt River Project could use the information to avoid disturbing endangered species when it located transmission lines and other facilities.

"He got it — that this was actually a very useful tool," Campbell said. SRP became the first entity in the state to put the database to practical use.

Northwest of Phoenix is a lush oasis, one of the last spots in central Arizona with a stretch of free-flowing river. The Nature

Conservancy established the 660-acre Hassayampa River Preserve here in 1987 to protect a rare strip of green in the desert. It includes Palm Lake, a spring-fed pond and marsh that attract some 280 species of local and migrating birds.

As the preserve was about to open, Campbell chatted with Jack about growing up in Arizona and loving its unique landscape. Why not have Jack speak at the rollout ceremony? he thought. The choice was risky. Conservation groups were lambasting SRP over emissions from coal-fired plants. A talk from its top executive could be very badly received.

More than a hundred people came to the opening ceremony. Jack spoke to them in very personal terms of how much it meant to him as a young boy to hike, picnic, and enjoy Arizona's natural areas.

Quoting from naturalist Henry David Thoreau's *Walden*, he "gave one of the most persuasive talks about the need for protecting Arizona heritage of anyone I've heard," Campbell said. He didn't shrink from talking about the industry's impact on the natural world and described SRP's efforts and challenges in emissions control.

At one point, Campbell was startled to realize that Jack was tearing up.

Instead of the protests Campbell feared, Jack was mobbed afterwards by people applauding his comments. Jack wasn't a charismatic speaker, but he captured audiences because he spoke with genuine sincerity and found something to say beyond platitudes. The Nature Conservancy's regional director, soon to become president of the organization, had flown in for the event and said he'd never seen anyone who had an audience "eating out of his hand" like that.

In 1992, Jack joined the board of trustees of the Nature Conservancy in Arizona, serving as chair from 1993 to 1995. When he stepped down in 1999, he stayed involved through the Chair's Council, a group of past trustees, until it disbanded in June 2009.

The one thing Jack didn't manage to do was look like a real outdoorsman. While others were wearing jeans and boots, he would go on field trips in slacks and black shoes.

◇ ◇ ◇

The election of 2000, however, put a chill in Jack's relations with the Nature Conservancy. The group joined other environmentalists in defeating Proposition 100, the land-preservation measure that had come out of his work as chair of the Growing Smarter Commission.

A year or so later, the Nature Conservancy went through some upheavals on the board, partly because the new state director, Pat Graham, was setting up his office in Phoenix instead of Tucson. He set out to mend fences and went to meet Jack. "I found him to be an incredibly gracious person," Graham said. "He talked a little about his philosophy: 'This community has been really good to me, and now it's my turn to give back.'"4

Jack wasn't content with following that principle; he was trying to spread it. "He sought people out who had been successful in the community and encouraged them that it was their turn to give back," Graham said.

Jack Pfister with Representatives Bob Stump, Jay Rhodes, and Jon Kyl, in 1987.

In 2003, the *Washington Post* ran a series of blistering articles about the Nature Conservancy's close ties to business and a series of questionable moves it had taken. One of the conservancy's key tools for protecting land from development, conservation easements, was also under attack. The idea was to compensate owners for giving up the opportunity to develop their property. But some clever developers had figured out how to profit from "protecting" places like golf courses.

The Senate Finance Committee was planning hearings, and Arizona Sen. Jon Kyl was a leading member. Graham didn't know Kyl and had no access to him. But the national office of the Nature Conservancy was pressing state directors to talk to senators on the committee about the value of its work.

Learning of Jack's longstanding relation with Kyl, Graham called him for help, and "he immediately agreed to get me in." For the meeting, Jack decided it would be best to take the lead himself and use Graham as a resource. It was the right move.

"There's nothing I could have said would have been as valuable as having Jack Pfister in the room to stand up and say this is a good organization," Graham said.

That issue was over in a flash. Then the conversation moved on to forest management. Everyone knew the West's overgrown forests had to be thinned out or the number of catastrophic wildfires would keep growing. Arizona had just seen massive destruction from the Rodeo-Chediski fire, which incinerated nearly half a million acres in 2002.

A meeting that was supposed to last ten minutes went on for forty-five. The three men tossed around the problem. Finding a solution, Kyl said, came down to two things: economics and trust.

"This seems pretty self-evident now," Graham said, "but it wasn't then. We needed to find a way to get business back in the forest that allowed them to make a profit on small-diameter trees. We needed to do it in a way that rebuilt the trust that had broken down between the environmental groups and the U.S. Forest Service and other stakeholders."

Kyl's observation crystalized the challenge, setting the Nature Conservancy on a course of forest management that might not have happened otherwise.

People think of helping an organization in terms of money and time. But, as Graham found, it can be equally valuable to offer your reputation and make connections with people who can make a difference.

When Jack took him to meet with Kyl, Graham said, "We went in with one purpose, we came out with another that was way bigger. We're still getting dividends from that today."

The national debate over illegal immigration was turning venomous in 2003. It often degenerated into the kind of "us" vs. "them" struggle that Jack tried to avoid, whether the issue was labor relations, tribal water rights, or the Martin Luther King holiday. When he talked about respecting differences and listening to others, it wasn't rhetoric but a conviction rooted in his soul.

Jack's openness could have its flaws, in former Phoenix Mayor Goddard's view: "Some of the people he was willing to embrace, I really think are very bad people. Yet because they are part of the community, he was very inclusive."[5]

Jack had invested a lot of time over the years in building bridges to span racial, political, religious, and ethnic differences. Now he saw high-profile politicians taking a torch to them.

Arizona Rep. J.D. Hayworth wrote an incendiary op-ed piece about illegal immigration, which appeared in the *Arizona Republic* in November 2003. Jack was appalled to read toxic words like "greed-driven orgy" and "hordes of illegal workers" and inflammatory assertions dressed up as facts.

He responded with his own op-ed piece two weeks later. Jack spent only a few sentences taking Hayworth to task. Instead, he argued that progress depended on "rational and civil dialogue among people of goodwill."

Victory Together's success on Martin Luther King Jr. Day had shown how Arizonans could come together to deal with divisive issues. But now, Jack lamented, "The rhetoric over immigration issues suggests that we have dissipated the civility and good will that 'Victory Together' worked so hard to achieve."[6]

He urged Arizonans to "join together again to work with our political leaders to find commonsense solutions to the vexing social problems of immigration." It was the Pfister formula. But it never happened. Without someone of Jack's stature, inclusiveness and endless capacity for dialogue, the issue continued to polarize Arizona.

Jack's idealism, though, shaped several generations of leaders after him. Entrepreneur John Fees was a rather-cocky student representative when Jack was on the Board of Regents. But Jack spotted potential that he was willing to foster. He coached Fees through a series of start-up companies, using his performance curve as guide.[7]

They talked about values and goals. Jack had concluded that it was important to have a personal mission statement and objectives and periodically evaluate your performance. His own basic mission, he told Fees, was "to live with integrity and to make a difference in the lives of others." He aimed to read two books a month and expand his circle of friends by adding four or five new friends a year.

Jack also set an objective for political involvement, to be "a concerned and an informed citizen, involved in the political process to ensure my voice is heard and my vote counted." If he had gotten a grade on that point, it would have been A-plus.

Jack's supposed retirement didn't leave much time for leisure activities. He'd had a separate woodshop since the house was remodeled years ago, but his carpentry tools were gathering dust. He had whacked balls for years as a mediocre golfer, gamely participating in SRP tournaments, but gave up when he realized he wasn't getting any better. While bird-watching fascinated Jack and Pat for a while, their interest faded. They didn't take major trips now, but they enjoyed traveling to Ashland, Oregon, for the annual Shakespeare festival.

With his passion for reading, Jack had joined a book group shortly after retiring from SRP. The reading list was almost a "great books" course — perfect for Jack's unflagging devotion to self-education. The group made its way through Shakespeare in the 1990s and moved on to ancient Greeks and the Bible.

Phoenix attorney Paul Eckstein, a founding member of the group, had met Jack professionally some three decades earlier. "I would describe him as having an engineer's mind and a humanist's heart," Eckstein said.[8]

Jack's comments were very different from those of others in the group, which was heavily weighted toward lawyers and professors. He was always looking for lessons in literature that could be applied in business and civic life. "More than anyone in our group, he was making those concrete connections," Eckstein said.

Jack would take his turn running a session and hosting it at his house. Pat would put out some drinks and nibbles and make small talk beforehand. But she never took part in the group and quietly disappeared as soon as the discussion started.

Jack had a different way of handling a book-group meeting, too His presentations were highly organized, and he liked to break into smaller groups to talk about particular questions.

While he spoke with ease during the discussions, Jack was oddly inept at reading aloud. When his turn came to read a passage, he spoke slowly and haltingly in a flat voice. It was obvious why friends remembered that he was picked for the nonspeaking parts in high school plays.

The Venice of Arizona

The Phoenix metro area has 181 miles of canals, more than Amsterdam and Venice put together. But instead of becoming a showcase, they were hidden behind buildings as the Valley grew, more like alleys than waterways. That's the way Salt River Project wanted it when Jack began working there in 1970. While the federal government owns much of the Valley's canal system, SRP manages most of it, except for the Central Arizona Project. SRP traditionally viewed the canals as tools for moving water, not urban amenities,. The public needed to be kept away from them.

Meanwhile, Valley residents saw how other cities were turning their waterfronts into magnets for entertainment, tourism, and development. A move began in the 1980s to beautify and take advantage of the canals in similar ways. In 1988, impressed by San Antonio's River Walk, Jack enthusiastically embraced the vision of development along canal banks.[1] The enthusiasm wasn't universal at the utility he led. "SRP viewed the canals as a pipeline," Jack said in an interview years later. "Re-imagining how the canals can be used has been a painful process for SRP."[2]

Nevertheless, with Jack's support, SRP began opening the way to public access. In the late 1980s it wrote guidelines for allowing multiple uses alongside the water delivery system. Around the same time, Scottsdale did a canal-banks study, which spurred the Junior League of Arizona to push for region-wide design and use principles. SRP contributed $25,000 to help fund that project, which was done through Arizona State University. (An additional $40,000 came from National Endowment for the Arts and $30,500 from seven local cities and towns.) The canal system "has the potential for being an incredible asset for our community, indeed reshaping its future for the 21st century," Jack said when the plan was released in 1990.[3]

Bit by bit, projects began taking advantage of that asset. Phoenix did several demonstration projects with art and pedestrian-friendly features along the canal banks. In 2004, Scottsdale broke ground on the massive mixed-use Waterfront development flanking a downtown stretch of canal.

Jack became a big fan of "canalscape" — creating hubs of activity where canals intersect major streets, with the banks becoming bike/pedestrian routes — after ASU Professor Nan Ellin described the concept in a March 2008 op-ed piece in the *Arizona Republic*. "He was extremely supportive," Ellin said.[4] As more of SRP's leadership came from urban interests, he assured her, the support for canalscape would grow. In the meantime, he encouraged her to meet with those board members who already had an urban perspective.

"He was a kind soul and great strategist," recalled Ellin, who later became dean of the College of Architecture, Planning and Public Affairs at University of Texas at Arlington. He advised her to focus on creating one outstanding example of a canalscape. "Envy is the great motivator," he said. Once developers and residents of other neighborhoods saw the project's success, they would be eager to recreate it.

ASU published an exploration of the concept, *Canalscape: An Authentic & Sustainable Desert Urbanism for Metro Phoenix*, in conjunction with a design competition and an exhibit at the ASU Art Museum and with support from SRP. It came out in November 2009, four months after Jack's death. The publication is dedicated to him as a "*pillar and light* — supporting and inspiring a better Metro Phoenix for all."

The Common Core of Decency

Jack was always uncomfortable about getting awards and honors for his work. He never wanted anyone to think he was after public recognition. For any success, he liked to spread the credit around. In part, that was a sign of his kindness and genuine appreciation of people's efforts. It was also a canny strategy to make people eager to pitch in the next time he asked them to help on a project.

But as the accomplishments in his life added up, he received a steady stream of awards.

In 2003, he was named a Historymaker. For someone so fascinated by Arizona's heritage, it was a special honor to receive that award from the Historical League, a support arm of the Arizona Historical Society. The league's biography of Jack quoted his advice to students at ASU: "Work hard to be happy. It's amazing what you can accomplish if you don't care who gets the credit."

In 2006, the law faculty at his alma mater, the University of Arizona, voted to give him the Distinguished Alumnus Convocation Award. The honor was given at the May graduation ceremony. What pleased and impressed Jack was the number of minority students he saw — which was certainly due in part to efforts by him and his fellow regents to widen the mix of Arizona university students.

In a thank-you note after the gathering, Jack wrote, "Perhaps the most gratifying part of the event was to observe the diversity of the graduating class. More diversity crossed the stage in the first five minutes of the Convocation than were in my entire law class."

Somehow in his supercharged schedule over the years, Jack had found time to serve on the founding board of directors of the

Phoenix Public Library Foundation. He was honored at a reception and fundraiser in 2007.

In an area dear to his heart, Jack received the Polly Rosenbaum Award in February 2009 for his role in rescuing Arizona's historical records. For years, the Legislature had ignored the crumbling and inadequate space where the state archives were stored, where leaks and heat threatened priceless photos and documents. They included the original state constitution and Wyatt Earp's extradition papers after the legendary O.K. Corral shootout.

Jack joined the crusade to show lawmakers that the records weren't just artifacts but indispensable economic and legal tools for current problems, such as settling land disputes. He put his political skills and contacts to work in the long, ultimately successful campaign to persuade the Legislature to fund a state-of-the-art home for the archives.

Jack had bought the Prescott home that his aunt and uncle had owned. It became a getaway and holiday spot for the family.

Built in 1900, the green-trimmed white house is one of Prescott's historic properties. Over the years, the detached garage had gotten more and more rundown. Jack finally found the time to get it fixed up — and received one of the annual restoration awards from the Prescott Preservation Commission in 2006.

Jack and Pat reached the milestone fiftieth wedding anniversary in 2006. Their children, Suzanne and Scott, threw a luncheon for them at the University Club in central Phoenix. Some seventy people came, including far-flung family, Sen. Jon Kyl, and old Salt River Project colleagues.

"To me, what held them together was that they both came from very modest means and they never lost those basic values," Suzanne said. "Money did not mean a lot to either of them."[1] They had a comfortable relationship, leaving each other space to pursue personal interests. Pat had her own causes, particularly the Heard Museum, where she was a longtime volunteer in supporting its commitment to American Indian art and history.

Pat and Jack at the Arizona Historymakers dinner, 2003.

The couple "didn't fight at all, which was good growing up," Suzanne said. "They were very, very compatible."

Pat was essential to Jack's success, said Loretta Avent, his partner on Harmony Alliance and a longtime family friend. "Jack could not have been the man he was without the wife he had," she said. "Very few women could ever give up that much time of their spouse on behalf of 'the People.'"[2]

They had gone through health crises together. Pat had a stroke in the late 1990s. Her whole left side was paralyzed when Jack rushed her to the hospital. A clot-busting drug worked a virtual miracle. Within an hour, she had nothing but some drooping of her face. She recovered so well on her own that she didn't need physical therapy.

The family suspected, though, that she suffered smaller strokes later on that caused some memory lapses. But if there were problems, Jack covered for her. He would take care of the things she'd handled if necessary. Whatever issues there were, no one knew. They kept parts of their lives very private.

Jack's advice to students and those he mentored usually included a version of the old saying, "No one on their deathbed ever said, 'I wish I'd worked more.'"

He didn't follow his own counsel. In a newspaper feature about wisdom gained in life, Jack was asked about regrets. What was the one thing he would have changed? "I would have spent more time with my family while I was working," he answered. "I did not have a balanced life." [3]

Finally, though, he'd balanced his commitments enough to focus on the history he wanted to write. In August 2007, the "Seen & Heard" column of the *Arizona Republic* noted that "retired Salt River Project executive Jack Pfister has taken up a new hobby: writing." Jack was concentrating on Burton Barr, the Republican dealmaker who was majority leader in the Arizona House for twenty years. But the next year, the project had a sad setback. His writing partner on the biography, retired Arizona State University professor Brent Brown, died in May 2008 at the age of 66. The book wouldn't be written in Jack's lifetime. The family passed his research on to ASU associate professor Philip VanderMeer, whose biography, *Burton Barr: Political Leadership and the Transformation of Arizona*, was published by the University of Arizona Press in 2014.

Jack was still quick to respond to the dozens, or maybe hundreds, of people who called him for advice. "Many of us thought of him as indestructible," said former Phoenix Mayor Terry Goddard. "The fact that he could have serious health issues never crossed our minds. He was Jack, he seemed like a rock, a fortress."[4]

But the years of smoking had lit a slow fuse. Jack had optimistically figured that so many years without a cigarette would put him out of danger. He and Pat were shocked when he was diagnosed with lung cancer in mid-2007.

The announcement didn't rattle Suzanne as much. She talked about disease every day as vice president of external affairs at St. Joseph's Hospital and Medical Center, and she knew it had an excellent treatment program for lung cancer. Her father's case was caught early. The odds were good that he could beat it.

A day or so after the diagnosis, Suzanne got a couple of books about cancer and gave them to Jack to look at when he felt ready. "He said it was the best thing that happened," she said, "because he jumped into fact-finding mode. And that was a way to sort of work through it."[5] As it did for many patients, information gave him a sense of control.

Jack had surgery that summer and then went through chemotherapy and radiation. The health threat pulled him closer to his brother, Tad, and they spoke on the phone every day.

Jack and his brother, Tad

Jack wasn't going to let his experience go to waste. He reached out to share with others. When he heard that his friend Saul Diskin had cancer, he immediately got in touch.[6] Jack offered sympathy, but he didn't linger on the point. "He was a man of action," Diskin recalled in 2013, a year before losing his own battle with the disease. "He sent me a book on how to choose doctors when you're diagnosed with cancer. It was an extremely good how-to book."

They had lunch from time to time. "We'd compare notes about medical treatment," Diskin said. "It was more clinical than it was moments of exchanging feelings, which, by the way, I appreciated. When there's a problem to solve, that's where you have to put your energies."

Although few people knew it, Jack had been dealing with serious medical issues for years. He was diabetic. And sometime after leaving SRP, he was diagnosed with rheumatoid arthritis.

It was the one time he couldn't seem to cope. "I remember vividly him blowing up, because it was so unusual," Suzanne recalled. "He never got mad. We were at dinner one night at a restaurant. And he got really, really sort of inappropriately angry."

As she talked to her father, it was clear that he was struggling with the diagnosis and she realized why. Jack's Aunt Bet had been terribly crippled by rheumatoid arthritis. Aspirin was the only treatment at the time, and it destroyed her stomach, leaving her virtually unable to eat and in terrible pain. Given genetics, Jack figured he was doomed to the same suffering.

The gloom went on for a month or so with no signs of letting up. "This is not a guy who would go to an arthritis support group," Suzanne said. So she called Carolyn Allen, a state senator and close family friend, who had rheumatoid arthritis herself. Suzanne asked her to speak to Jack and be brutally honest. It wasn't likely to be otherwise. Allen was known as a no-nonsense legislator, who put constituents above ideology and never minced words.

She went to lunch with Jack and found him hurt and angry. After a life of hard work, well-earned respect, and civic service, he couldn't believe this was happening. Why him? Why now? Why was he being punished? "Get over it, Jack," Allen responded. "You're no better than anyone else. Suck it up. You're going to live with it, because you don't have any other options."[7]

She understood Jack's fear of being crippled. When she was diagnosed twenty years earlier, the doctor had said she'd end up in a wheelchair. "And I decided, piss on you, I'm not going to," she said. "I'm not in a wheelchair, Jack. And you're not going to be, either, if you do what they tell you."

The standard treatment was a weekly dose of methotrexate. Jack was amazed that Allen gave herself the shot — she brought the needle to lunch to show him how small it was. "I do mine every Sunday," she said. "You pick a day. You take your shot and shut up about it."

Jack listened politely, although his face flushed a few times. "He was too much of a gentleman, I think, to flare up at me," Allen said. And then he got caught up in questions about side effects, testing, logistics. His natural curiosity took over.

Allen had snapped him out of his funk. He knew the lunch wasn't happenstance. But all he said to Suzanne was, "I know you orchestrated this. Thank you for making this happen."

Treatment had put Jack's cancer into remission. But in less than a year, tests showed it was back. He went through three biopsies in 2009 as doctors tried to confirm the diagnosis. At one point, they took fifty scrapes of tissue, but still couldn't get a clear result. Jack was given the option of waiting — if it was indeed cancer, the evidence would be clearer over time. But he was adamant: "I want to know." And finally a test had the dreaded definitive answer: Yes.

Jack went through another round of chemotherapy that summer and had more scheduled for late July.

He kept up his breakfast and lunch meetings with the vast number of people he'd mentored and befriended over the years. Cathy Eden was going to Ireland that July, but she squeezed in breakfast with Jack before leaving.[8]

They were in the habit of meeting every couple of months and talking about public policy. Jack loved political gossip — not personal stories, but the tales of intrigue and the real explanation for dead legislation or a sudden job shuffle. When he heard a secret, though, he never revealed it. "He held everyone's confidences," said Eden, whose career included serving as Coconino County manager, a legislator, and director of the state departments of administration and health.

Whenever she needed help, she knew she could count on Jack. She'd turned to Jack at a particularly critical moment in 1989. She was director of the Department of Administration, overseeing the nuts and bolts of state government. State employees hadn't had a raise in more than five years. Cities and counties, which had far better pay scales, were picking off staff at an alarming rate. Eden had to ask the Legislature to increase state salaries, and she called Jack for a hand in making the business argument.

"We put together a lot of data, graphs, and charts to explain why it was important," she said. Then Jack told legislators why businesses cared: They had to deal with state agencies, and they wanted the staff to be bright and competent — the kind of people who are lost without adequate pay. "He went after the Republicans pretty hard," Eden said. The raise was approved.

At that breakfast in July, Jack was coughing and obviously not doing well. "Tell me what's going on," Eden said.

"The cancer's come back," he replied. But he was upbeat, saying the doctors thought it was treatable.

Jack had been more somber when he told attorney Ernest Calderón that the cancer had recurred.[9] As usual, the conversation was over breakfast. They talked about his age. At seventy-five, Jack was lucky — a lot of men hadn't made it that far in life. They discussed how fortunate he was. He had access to excellent medical treatment when not everyone did.

Jack's nature was always to look for the positive, but this time, Calderón said, "I didn't sense the optimism."

The end came on July 20. Jack collapsed at home. Pat called 911, but the paramedics couldn't revive him.

The memorial service was held at the Orpheum Theatre in Phoenix — fittingly, a historic building. It was crowded with people who had been touched by Jack. Strangers came up to the family with stories of his profound influence on their lives. At one point in the memorial gathering, people were asked to raise their hands if they'd been mentored by Jack. The hall filled with raised hands.

Jack's impact on Arizona was perhaps best explained years earlier, in 1992, when he received the Human Relations Award from the

National Conference of Christians and Jews. Lattie Coor, president of ASU at the time, gave this assessment:

"Jack Pfister combines an uncommon ability at problem-solving with a boundless commitment to the well-being of this community and this state. By listening carefully to all who are around him, he finds the common core of decency by which a society advances itself and helps us all to make this a better place in which to live."[10]

Jack was introduced to the Arizona Historymakers crowd by retired ASU President Lattie Coor in 2003.

A Note on Sources

Much of this story of Jack Pfister's life is drawn from news articles, Salt River Project publications, organizational newsletters, government reports, and more than eighty interviews. Jack's life was so varied and so crammed with complex and technical issues that virtually every paragraph could have been footnoted at least once. As explained in the endnotes, I've chosen for the sake of readability to keep notes largely limited to identifying the source of quotations.

Jack Pfister is a biographer's dream. With his passion for history, he knew the importance of hanging onto source material. He assembled several notebooks of family history and filled notebooks with clippings and other material about Salt River Project, his subsequent career, and the many issues he was involved with. He kept his correspondence and agenda books. He donated an extensive collection of papers and books to Arizona State University, although much of it concerns historical events before his time.

The Arizona Memory Project, azmemory.azlibrary.gov, offers access to a number of the reports that Jack worked on.

Jack recorded several oral histories. The most extensive was in 1991, an invaluable interview conducted by Karen L. Smith as a historical record for Salt River Project. SRP continued to do oral histories, and parts of some provided helpful background for this biography. The Central Arizona Project has assembled an impressive collection of oral histories of people involved in the project, including those who fought it. The edited transcripts, including Jack's, are available on-line at www.cap-az.com/documents/about/oral-histories/Interview_with_Jack_Pfister.pdf. Jack gave a broader perspective in an oral history for the Historical League, Inc., when he was named an Arizona Historymaker for 2003. A transcript of the interview, conducted by Pam Stevenson and videotaped by Bill Stevenson on July 23, 2002, is linked to a biography on the website: www.historicalleague.org.

Jack established a quarterly newsletter, "Arizona Waterline," in 1982 as a forum for discussing water issues. Published by SRP and intended to last two years, the publication went on for six. Edited by Athia L. Hardt and later collected in a book, *Arizona Waterline* (Phoenix: Salt River Project, 1989), the articles are a valuable record of the challenges and responses of the 1980s. As of 2015, it was still easy to track down inexpensive copies of the book.

Since links may change over time and search engines make it easy to track down a report, I've included web addresses in only a limited number of cases.

Family Roots: Jack conducted extensive research into his family history, which he assembled in two notebooks. The material includes copies of death certificates, letters, newspaper articles, family trees, a narrative written by Jack, and a mimeographed family history written by his uncle, Leon F. Kneipp, dated August 11, 1947.

Arizona History: Many excellent books cover Arizona and Phoenix history. Thomas E. Sheridan, *Arizona: A History*, (Tucson: University of Arizona Press, 2012) offers an excellent overview.

Central Arizona Project: Background on the project, particularly before construction, is in Jennifer E. Zuniga, "The Central Arizona Project," Bureau of Reclamation, 2000. A detailed discussion of the Carter hit list is in "Water Projects Compromise Reached," *CQ Almanac 1977*, 33rd ed., 650-59. Washington, DC: Congressional Quarterly, 1978.

Phoenix 40: A brief history is on the website of its successor organization, Greater Phoenix Leadership, www.gplinc.org.

Floods: A short history of the SRP canal system and the September 1970 flooding is included in the Arizona Supreme Court decision in Salt River Water Users Assn. v. Giglio. Two federal reports analyze the huge 1980 floods: "The Disastrous Southern California and Central Arizona Floods, Flash Floods, and Mudslides of February

1980," Natural Disaster Survey Report NWS 81-1 (Silver Spring, Md.: National Weather Service, 1981) and "Floods of February 1980 in Southern California and Central Arizona," by B.N. Aldridge and R.J. Longfield, U.S. Geological Survey and E.H. Chin, National Oceanic and Atmospheric Administration, (Washington: U.S. Government Printing Office, 1991). Jack himself wrote an account in "The Day Stewart Mountain Dam Almost Failed," a conference paper prepared for the annual Arizona History Convention in April 2003; it is available in the Jack Pifster Papers, Arizona Collection, Arizona State University Libraries.

Three Mile Island Accident: The "Report of the President's Commission on the Accident at Three Mile Island" is available on-line. The U.S. Nuclear Regulatory Commission produced a handy sheet, "Backgrounder on the Three Mile Island Accident," available on its website.

SRP in Egypt: For a contemporary analysis of the objectives and accomplishments of the 1977-1984 water development effort in Egypt, see the "Final Administrative Report, Egypt Water Use and Management Project," by E.V. Richardson, M.E. Quenemoen and H.R. Horsey, Consortium for International Development, April 1985.

Goundwater Management: For the impact of pumping, see "Land Subsidence and Earth Fissures in Arizona," by Steven Slaff (Tucson: Arizona Geological Survey, 1993) and "Land Subsidence and Earth Fissures in Arizona: Research and Informational Needs for Effective Risk Management," by Arizona Land Subsidence Group (Arizona Geological Survey report, 2007). Background on legal efforts to control pumping are in "Final Report, June 1980," Arizona Groundwater Management Study Commission, and in an article reprinted in *Arizona Waterline*, "Struggle to Get Act Frustrating, Rewarding," by A.J. Pfister and Rep. Larry Hawke, 1987.

Tribal Water Rights: A full and excellent discussion is in Daniel Killoren's 2011 Arizona State University doctoral dissertation, "American Indian Water Rights in Arizona; From Conflict to Settlement, 1950-2004." A useful summary is in "Appendix G: Indian

Water Rights Claims and Settlements," in the *Arizona Water Atlas*, Volume 1, available on the website www.azwater.gov.

Plan 6: The process is laid out in "Public Involvement Plan: Central Arizona Water Control Study," (Water and Power Resources Service, Arizona Projects Office, 1980). For a discussion of habitat concerns omitted from the Orme Dam draft environmental impact statement, see the July 2, 1976, letter from Laurie J. Vitt, who participated in part of the assessment, to Manuel Lopez Jr., regional director of the U.S. Department of Interior. http://archive.library. nau.edu/cdm/ref/collection/cpa/id/61157

Harmony Alliance Inc.: Loretta Avent kindly provided access to the organization's files, which were not publicly available as of 2015.

Efforts to Protect the San Pedro River: The full report from the panel Jack co-chaired, *A Ribbon of Life: An Agenda for Preserving Transboundary Migratory Bird Habitat on the Upper San Pedro River,* is available online. Unfortunately, the long-term fate of the San Pedro River remains in the balance, as both mining and residential development projects threaten to deplete the groundwater vital for its year-round flow.

ASU Research Park: A review of the history of research parks, as well as an analysis of those in Arizona, is in "Arizona Universities' Research Parks," a performance audit by the state auditor general, Report #97-20, November 1997, available on-line through the Arizona Memory Project.

Phoenix-area Canals: For proposals to make the canals part of the urban fabric, see "Metropolitan Canals: A Regional Design Framework," Michael Fifield, Madis Pihlak, Edward Cook, Sharon Southerland, (Tempe: College of Architecture and Environmental Design, Arizona State University, 1990) and "Canalscape: An Authentic and Sustainable Desert Urbanism for Metro Phoenix," Nan Ellin, project director, (Tempe, Arizona State University, 2009).

Notes

For readability, notes have been kept to a minimum, mostly to identify the source of quotations. When the source is a personal interview, only the first quotation is noted. Likewise, when the source is one of the major oral histories Jack did, only the first quotation is noted. Further information is in the Note on Sources.

Abbreviations

AR *Arizona Republic*

PG *Phoenix Gazette*

Pfister 1991 Oral history interviews conducted by Karen L. Smith in 1991

Introduction

[1] Jack Pfister, "Arizona Needs To Develop New Goals, Leaders," *Scottsdale Progress*, January 1, 1991.

[2] Jon Kyl interview, December 16, 2013.

Prescott Roots

[1] Description of performances from undated clippings in Jack Pfister's personal notebooks.

[2] Jack Pfister, Historymakers oral interview, July 23, 2002.

[3] Jack Pfister, 1991.

[4] Inez McDonnell, letter to Leon Kneipp, circa January 4, 1937.

[5] Elisabeth Ruffner interview, May 11, 2013.

[6] Tad Pfister interview, June 1-2, 2013.

A World of Learning

[1] Pfister, 1991.

[2] Jack Pfister, memoir written for centennial of Washington School.

[3] Mack Wiltcher interview, August 11, 2013.

[4] Pfister, 2002.

[5] Jack Pfister, undated interview.

[6] "Early Life Taught Jack Pfister Value of Hard Work, Concern for Others," *Ed Cetera*, Fall 1996.

[7] Pfister, 2002.

[8] Pat Pfister interview, February 9, 2014.

The Road to Salt River Project

[1] Pfister, 1991.

[2] Jack DeBolske interview, January 23, 2014.

[3] Tad Pfister, 2013.

[4] Suzanne Pfister interviews, May 11-12, June 2-3, 2013.

[5] Terry Goddard interview, May 14, 2014.

Coal and Clean Air

[1] Leroy Michael interview, August 27, 2013.

[2] Al Qöyawayma interview, February 28, 2014.

[3] Jack Pfister, oral history for Central Arizona Project, May 23, 2005.

Taming Wildcats

[1] Scott Pfister interview, June 24, 2013.

[2] A.J. Pfister to L.B. Jolly, vice president of Mountain Bell, December 22, 1970.

[3] Suzanne Pfister, 2013.

[4] Bob Mason interview, July 2, 2013, January 21, 2014.

[5] Stanley Lubin interview, February 25, 2014.

[6] Source interviewed by author, 2013.

[7] Michael, 2013.

[8] Pfister, 2002.

At the Top of the Ladder

[1] Pfister, 1991.

[2] "Why Pay Utilities' Costs," letter to the editor from Kenneth F. Biehl, AR, August 18, 1976.

[3] Richard Silverman interview, June 3, 2013.

[4] Grant E. Smith, "SRP Boosts Electric Bills by 26.7 Pct.," AR, December 21, 1977.

[5] Bill Shover interview, November 13, 2013.

[6] Comment from interviewee to author.

[7] Bill Davis interview, March 11, 2014.

[8] Sid Wilson interview, November 18, 2013.

[9] Suzanne Pfister, 2013.

[10] "Association Forms Political Action Committee," *Pulse,* August 19, 1976.

[11] Athia l. Hardt, "SRP Agency Has Donated $1,900 Thus Far in Races," AR, undated clip in Pfister files.

[12] A.J. Pfister to Ray, Stan and Tony, December 10, 1976.

[13] "General Manager Comments on Economic Pickups, Slowdowns," clipping from SRP publication, n.d.

[14] "Project Initiates Power Saver Service to Help Customers Conserve Energy," *Pulse,* April 12, 1977.

[15] Tom Kuhn, "SRP to Make Loans for Insulation Work," AR, April 13, 1977.

[16] "'40' Sets sights on Community Concern, Hope," clipping from unidentified newspaper, March 3, 1976.

[17] Dennis Mitchem interview, July 31, 2013.

[18] Lois Boyles, "Ever Alert, Phoenix 40 Finds Few Issues to Sink Teeth Into," PG, August 4, 1978.

[19] Laurie J. Vitt, letter to Manuel Lopez Jr., Regional director of the Bureau of Reclamation, July 2, 1976.

[20] Lou Hiner, "State Leaders Defend CAP at Hearing," PG, March 21, 1977.

[21] Cecil Andrus, oral history for Central Arizona Project, June 6, 2005.

fister's Air Force

[1] Pfister, 1991.

[2] "GM Talks About Negotiations, Wages and the Union," *Info,* October 17, 1977.

[3] "Electricians Walk Out at SRP Plants," AR, January 11, 1978.

[4] "Tribe Claims SRP Magnifies Strike Violence," AR, January 15, 1978.

[5] "Union Boss Denies Harassment of SRP," AR, January 18, 1978.

[6] "Electricians Walk Out at SRP Plants," AR, January 11, 1978.

Deluge in the Desert

[1] Pfister, 1991.

[2] Earl Zarbin, "Worse Floods Possible in '79, SRP Reports," AR, April 11, 1978.

[3] "Flood Loss Blamed on State and SRP by Audubon chief," AR, March 8, 1978.

[4] Zarbin, "Worse Floods Possible."

On the Edge of Disaster

[1] Pfister, 1991.

[2] Tom Sands interview, June 25, 2013.

[3] Earl Zarbin, "Panel Told of Twin Flood-Control Problems," AR, February 16, 1979.

[4] "Problems Seen with 25 Dams," *Spokesman-Review,* October 20, 1977.

[5] Tom Fitzpatrick, "Weekend with Babbitt: Nuclear Waste to Floods," AR, February 18, 1980.

[6] Andrew Hurwitz interview, November 26, 2013.

[7] C.A. Howlett interview, November 15, 2013.

[8] "Safety—First Consideration at Stewart Mt.," *Pulse,* February 21, 1980.

[9] Don Harris and Jack Kowalec, "500-Year Flood Feared Possible," AR, February 16, 1980.

[10] Fitzpatrick, "Weekend with Babbitt."

[11] Jerry Seper, "Babbitt: Be Prepared for the "Unthinkable," AR, February 16, 1980.

[12] John Kolbe, "SRP Blames Weather for Valley Flooding," PG, February 23, 1980.

[13] Rebecca Watral, "Pfister Explains Flood Damage Causes," *Real Estate Press,* May 1980.

Nuclear Power on Trial

[1] Bruce Babbitt interview, November 16, 2013.

[2] John D. Selby, chairman of Consumers Power Company, "The Electric Industry's Response to Current Events," Outlook for Nuclear Power: Presentations at the Technical Session of the Annual Meeting, National Academy of Engineering, November 1, 1979.

[3] Claire Sargent interview, May 19, 2014.

[4] Mark Bonsall interview, April 2, 2014.

Reaching Out to the Pyramids

[1] Ed Kirdar interview, May 17, 2013.

[2] "Don't Believe All You Read in the Papers," Q&A with Jack Pfister, *Cross Currents,* May 20, 1985.

[3] "Seminar on Irrigation Techniques," *The Egyptian Gazette,* September 24, 1985. "Symposium on Irrigation and Drainage," *The Middle East Times,* September 29-October 5, 1985.

[4] Paul Ahler interview, August 17, 2014.

Pumping Arizona Dry

[1] Babbitt, 2013.

[2] Kyl, 2013.

[3] Pfister, 1991.

[4] DeBolske, 2013.

[5] Cited in A.J. Pfister, "Looking Ahead," 1989, in *Arizona Waterline,* ed. Athia L. Hardt (Phoenix: Salt River Project, n.d.).

[6] Anthony Sommer and Jack Lavelle, "Water Quality Legislation Needs Only Final Touches," PG, April 18, 1986.

[7] Kathleen Stanton, "Water-Quality Package Completed; Is Praised as Boon for Environment," AR, April 19, 1986.

The Tribes Are Thirsty

[1] A.J. Pfister and Karen Smith, "Indian Water Rights — the Conflict," 1984, in *Arizona Waterline.*

[2] Pfister, 1991.

[3] A.J. Pfister, "Resolution of Indian Water Claims," 1984, in *Arizona Waterline.*

[4] Richard Wilks interview, August 15, 2014.

[5] Daniel Killoren, "American Indian Water Rights in Arizona: From Conflict to Settlement, 1950-2004," PhD diss., Arizona State University, 2011.

[6] Michael Clinton interview, August 21, 2014.

[7] Deborah Laake, "Pfister Famine: With the SRP Chief's Retirement, We May Be Tapped Out on Skilled Negotiators," *Phoenix New Times,* November 14, 1990.

The Mathematics of Tuition

[1] Babbitt, 2013.

[2] Joel Nilsson, "Committee Endorses 2 as Regents; Confirmation by Senate Expected," AR, February 11, 1982

[3] Jane Erikson, "Regents Reject Fee Hike Asked by Legislators," *Tempe Daily News,* March 11, 1983.

[4] "Surcharge on Tuitions Wins Regents Approval," PG, April 16, 1983.

[5] Richard de Uriarte, "Newsmaker," PG, April 16, 1983.

[6] Richard de Uriarte, "New Entrance Rules To Require Geometry," PG, May 21, 1983.

[7] Dennis Russell, "Regents Supportive," *Glendale Star,* November 23, 1983.

[8] "'Sense of Realism': Regents Under Pressure to Find Equitable Tuition Policy," *Tempe Daily News,* January 30, 1984.

[9] Jack Lavelle, "Lawmakers Table Tuition Bill," PG, May 28, 1984.

[10] Ibid.

[11] Chip Warren, "In-state Tuition May Soon Pass $1,000," *Arizona Daily Star*, September 9, 1985.

[12] "Regents Okay Record Tuition Hike," *Arizona Capitol Times*, November 13, 1986.

[3] *State Press,* November 14, 1985.

[14] Jacque G. Petchel, "Regents Panel Shies from Taking Stand on S. Africa," AR, September 9, 1985.

[15] Michael Murphy, "Colleges Told to Cut S. Africa Ties," PG, September 7, 1985.

Saving the Affordable Option

[1] Pfister, 2002.

[2] Mitchem, 2013.

[3] Paul Elsner, email to author, September 12, 2014.

Rough Riding for the Regents

[1] Sam Stanton and Mary Jo Pitzl, "Mecham Assails Critics of His Education Policy," AR, February 20, 1987.

[2] Transcript of exchange, Pfister private papers.

[3] Steve Yozwiak, "Apology Is Sought from Regents Chief," AR, February 24, 1987.

[4] Jack Pfister, to Rep. Bob Denny, March 2, 1987, with partial transcript attached.

[5] Jack Lavelle, "Phoenix 40 Opposes Tax Reduction," PG, March 7, 1987.

[6] Tom Spratt, "Business Leaders Oppose Cut in Sales Tax," PG, March 18, 1987.

[7] Tom Spratt, "Regents Approve Massive Study of Universities," PG, March 21, 1987.

[8] Sam Stanton, "Governor, Regents Vie Over Study," AR, May 2, 1987.

[9] Glenn Rabinowitz, "Board Working To Improve State Education, Regents Says," *Glendale Star*, June 12, 1987.

[10] C.T. Revere, "Regents' Waste Hunt Fell Short, Critics Say," *Arizona Daily Star*, August 15, 1988.

Touchdown Season

[1] Richard de Uriarte, "Senate Committee OKs Regents Appointments," PG, February 10, 1982.

[2] Herman Chanen interview, April 22, 2014.

[3] Bill Hogan interview, April 22, 2014.

[4] Anthony Sommer, "Sky Box Plum Put in Cards' Gift Package," PG, December 17, 1987.

[5] Tom Spratt, "Alcohol Issue Called Vital to Cards Pact," PG, April 23, 1988.

[6] Mary Jo Pitzl, "Liquor Gets OK in Cards Skyboxes but Not in the Stands," AR, June 10, 1988.

[7] Paul G. Allvin, "Regents, Businesses Meet to Discuss Conflicts," *Arizona Daily Wildcat*, November 7, 1988.

[8] A.J. Pfister, "Chicano Conference Resignation Call Is Unwelcome," letter to the editor, AR, June 18, 1988

[9] Steve Nolen, "Nelson Quitting ASU Post," AR, August 20, 1988.

[10] Mike Rappoport interview, February 11, 2014.

[11] Ray Schultze, "Search Team Defends 1-Man 'List' for ASU Job," PG, May 26, 1989.

[12] Hurwitz, 2013.

[13] Mary Jo Pitzl, "Regents Back $84 Boost in Tuition," AR, December 3, 1988.

[14] Kristi Ellis, "Regents May Boost Student Financial Aid To Compensate for Federal Funding Cuts," PG, February 9, 1989.

[15] Jack Pfister, "Memorandum Regarding University of Arizona Presidency," November 19, 1990.

Getting to Plan 6

[1] Ben Cole, "Orme Options Are 'Bizarre,' Rhodes Says," AR, September 30, 1980.

[2] Babbitt, 2013.

[3] "SRP's Pfister Discusses Valley Flood Problems," *Times* of Fountain Hills, February 22, 1979; "SRP's View," Q&A with Jack Pfister, AR, April 1, 1979.

[4] Dr. Marty Rozelle interview, November 11, 2013.

[5] Mary A.M. Perry, "Governor's Panel Rejects Confluence Dam," AR, October 4, 1981.

[6] Mary A.M. Gindhart, "Water Panel Divided on Use of Local Funds to Speed CAP Work," AR, February 23, 1985.

[7] Mike McCloy, "SRP Chief Urges Plan 6 Be Kept Intact," PG, March 5, 1985.

[8] Mike McCloy, "Utility, Cities Resolve Dispute on Plan 6 Dams," PG, March 21, 1986.

[9] Carolina Butler interview, November 12, 2013.

[10] Robert Witzeman interview, October 25, 2013.

The Crucial Connection

[1] Tom Sands interview, June 25, 2013.

[2] Mary A.M. Perry, "SRP Wary of Being Valley's Pipeline for CAP Water," AR, February 28, 1983.

Strength from Diversity

[1] John Winters, "To Honor King: Babbitt Proclaims Holiday," AR, May 19, 1986.

[2] Dan Chu, "Arizona's Outspoken New Governor, Evan Mecham, Seems To Enjoy Diving Straight into Political Hot Water," *People,* August 24, 1987.

[3] Hardy Price, "Stevie Wonder Warns of Boycott If Mecham Cancels King Holiday," AR, November 15, 1986.

[4] Chip Warren, "ASU Students, Workers Ask Regents for King Holiday at All Universities," *Arizona Daily Star,* undated clip in Pfister papers.

[5] Mary Jo Pitzl, "Regents Approve Paid King Day at State Universities," AR, September 10, 1988.

[6] Don Harris, "Senators Pressed on King Day," AR, September 18, 1989.

[7] Dr. Warren H. Stewart Sr. interview, July 22, 2014.

[8] Pat Flannery, "State Soon Last Holdout on King Day," PG, May 16, 1991.

[9] Ben Winton, "Calm Approach Credited in King Day Vote," PG, November 12, 1992.

[10] Rev. Paul Eppinger interview, July 30, 2014.

[11] David Fritze, "Arizonans Bask in King Day Win," AR, November 5, 1992.

[12] Ben Winton, "Calm Approach Credited in King Day Vote," PG, November 12, 1992.

[13] Steve Roman interview, November 11, 2013.

[14] Sarah Auffret, "Gomez, Pfister Earn MLK Servant Leadership Awards, *ASU Insight,* January 13, 2006.

No Hat in the Ring

[1] Letter from James Derouin, August 5, 1987.

Getting Arizona in Harmony

[1] Ioanna Morfessis interview, July 20, 2014.

[2] Loretta Avent interview, February 21, 2014.

[3] Rabbi Robert Kravitz interview, April 1, 2014.

[4] Lisa Loo interview, May 28, 2014.

[5] David Fritze and Judy Nichols, "Minorities Describe Subtle 'Atmosphere' of Bias," AR, July 12, 1992.

[6] Amy Silverman, "Valley Leaders Organizing To Battle Racism," *Scottsdale Progress*, November 2, 1991.

[7] Duane K. Sheldon to Jack Pfister, November 25, 1991.

[8] Jack Pfister to Duane K. Sheldon, December 7, 1991.

[9] Harmony Alliance, Inc., Brief History, 1991-1994, Harmony files.

[10] Victor R. Beasley memo to A.J. Pfister and Loretta T. Avent, May 28 1991.

[11] Described in undated memo to members following March 4, 1992, meeting, Harmony files.

[12] David Fritze, "Little Changing at Clubs," AR, June 13, 1992.

[13] David Fritze, "First Blacks Join Valley Country Club," AR, September 23, 1992.

[14] José Cárdenas interview, May 28, 2014.

Adopted Grandkids

[1] Interview with Loretta, Bryant and Brittany Avent, August 26, 2014.

Peak Performance

[1] "Thoughts Shared by Jack Pfister: Jack — the 'Yoda' of Mentoring," slide presentation, compiled by John Fees, author's copy.

[2] Bonsall, 2014.

[3] Pat Denley, "PEP Videotape Shows SRP's Need for Improved Productivity," *Current*, December 13, 1982.

[4] Howlett, 2013.

[5] "Lecture Series Benefits SRP," *Current,* June 27, 1983.

[6] Rappoport, 2014.

[7] Ernest Calderón interview, May 21, 2014.

[8] Chris Herstam email to author, August 5, 2013.

[9] Pfister, 1991.

[10] Peter Aleshire, "SRP'S Pfister: 'A Major Force' in Water History," AR, October 18, 1990.

Protecting a Ribbon of Life

[1] Michelle Nijhuis, "Charting the Course of the San Pedro River," *High Country News,* April 12, 1999.

What a Blast

[1] "Don't Believe All You Read in the Papers," Q&A with Jack Pfister, *Cross Currents,* May 20, 1985.

[2] "New Media Will Improve Info Flow," *Cross Currents,* June 30, 1983.

[3] Karen Smith interview, July 2, 2013.

[4] Joel Garreau, *Edge City,* New York: Doubleday, 1991.

[5] RichardSilverman, 2013.

[6] "Those Rumors About Land Buys Were On-Target," Q&A with Jack Pfister, *Cross Currents,* June 30, 1983.

[7] Hal DeKeyser, "Utilityland: If You Think Legend City Was Thrilling, Just Wait," *Tempe Daily News,* June 19, 1983.

[8] John Staggs, "SRP Snafu Costs $400,000," AR, August 5, 1986.

[9] Bill Schrader interview, April 13, 2014.

[10] 1988 Phoenix Man and Woman of the Year nomination form, December 1, 1988, Pfister papers.

[11] Salt River Project annual report 1987-88.

[12] Guy Webster, "SRP Building 'Out of Control,'" AR, June 14, 1988.

Soaring Away

[1] Leslie Irwin, "SRP Board Approves Papago Project," PG, February 2, 1989.

[2] Smith, 2013.

[3] David Schwartz, "1,000 Might Lose Jobs Under SRP Cost Cutting," AR, March 15, 1989.

[4] "Soar Recommendations Raise Pointed Questions," Q&A, *Pulse,* March 23, 1989.

[5] David Schwartz, "SRP Apologizes Over Layoff Data," AR, March 29, 1989.

[6] Kirdar, 2013.

[7] Richard Silverman, 2013.

[8] David Schwartz, "Budget Ax Targets $475 Million at SRP," AR, September 6, 1990.

[9] Schrader, 2014.

[10] "Marketing, Economic Topics the Focus of Jan. 19 MCS," *Current Issues,* February 28, 1990.

[11] Dawn Gilbertson, "SRP Posts 1st Loss Since '47," PG, September 22, 1990.

[12] Guy Webster, "SRP Board OKs Cuts," AR, December 14, 1990.

[13] Dawn Gilbertson, "SRP to Replace Pfister with Longtime Employee," PG, January 7, 1991.

[14] Max Jarman, "Challenges Await SRP's New Leader," *Arizona Business Gazette,* February 1, 1991.

Government on a Diet

[1] J. Fife Symington III interview, March 25, 2015.

[2] "Political Troubles Hounded Leckie in Government," by Pat Flannery, AR, March 15, 1996.

[3] This quotation and the following ones from Jack are from a transcript of his March 28, 1994, interview with a Maricopa County Attorney's Office investigator, Pfister personal papers.

[4] "Verbatim Excerpts from the Report on the Project SLIM Contract," AR, July 24, 1994.

[5] Joel Nilsson, "Symington's Way of Handling Suspicious SLIM Bid: Ask George," AR, July 23, 1993.

[6] "Two Quit Project SLIM; Dispute Over Control Cited," *Arizona Capitol Times*, February 5, 1992; "SLIM Is Looking More Like a SHAM," editorial, *Scottsdale Progress,* February 11, 1992.

[3] Carol Sowers and Mary Jo Pitzl, "Head of Panel on Streamlined Spending Quits," AR, January 31, 1992.

[8] Carol Sowers, "Governor Loses 2nd Panelist," AR, February 1, 1992.

[9] Elliott Hibbs interview, February 7, 2014.

[10] Mary Jo Pitzl, "Cutting Government: State Regains SLIM's losses," AR, March 8, 1998.

[11] "Timeline: Project SLIM Indictments," AR, March 15, 1996.

[12] Jerry Kammer, "Yeoman: Leckie Leaked Bid Data," AR, May 17, 1997. The transcript was made public two days after the verdict.

[13] Coopers & Lybrand statement, September 20, 1996, Pfister personal papers.

The Smart Moves

[1] Jane Hull interview, May 8, 2014.

[2] Interview of Jack Pfister by Jeff Sprout, Topic: Students First/School Funding, March 29, 2001, unpublished.

[3] Maria Baier interview, May 26, 2014.

[4] Luther Propst interview, May 21, 2014.

[5] Kathleen Ingley, "Growing Smarter Gets Down to Work," AR, December 7, 1998.

[6] Steve Wilson, "Free Market Won't Stop Valley from Growing into a Mess,"AR, February 28, 1999.

[7] Ibid.

[8] Arlan Colton interview, August 5, 2014.

[9] Kathleen Ingley, "Growth Management Debate Wants You," AR, July 10, 1999.

[10] Gregor McGavin, "Drawing a Line in the San Tans," AR, July 25, 1999.

[11] Robert Robb, "Divergent Interests Stymie Consensus on Growing Smarter," AR, August 25, 1999.

[12] Dennis Wagner, "Smart-Growth Panels Oks List of Ideas, Splits on Trust-Land Plan," AR, September 1, 1999.

[13] Kathleen Ingley, "Hull Touts Growing Smarter Proposals," AR, November 2, 1999.

[14] Sandy Bahr interview, August 2014.

[15] "Hull Signs 'Historic' Growth-plan Bill," AR, February 22, 2000.

The Most Satisfying Job

[1] Joe Cayer interview, February 10, 2014.

[2] E.J. Montini, "Success 101: Rub My Back, I'll Rub Yours," AR, September 30, 1991.

[3] http://www.legacy.com/guestbooks/jack-pfister-condolences/ 130087742?#sthash.J8mEdP5y.dpbs

[4] Kim VanPelt interview, September 11, 2014.

[5] Lynette Summerill, "Pfister Retires From Public Affairs Post," *ASU Insights,* December 15, 2000.

[6] Lattie Coor interview, May 21, 2013.

[7] Nancy Neff interview, August 18, 2014.

[8] Pfister, 2002.

[9] Jack Pfister and Heather A. Okvat, editors, *The Beat the Odds Handbook,* Phoenix: Center for the Future of Arizona, 2010.

Research in the Park

[1] Hurwitz, 2013.

[2] John DeWitt, "Tenants, Subsidies Rise at Research Park," *Arizona Business Gazette,* December 24, 1992.

[3] Naaman Nickell, "ASU Research Park Starts to Take Off After Slow, Rocky Beginning," AR, December 21, 1995.

[4] Adam Kress, "ASU Research Park Begins 20th Year with $44M Boost," *Business Journal,* March 21, 2004.

The Torch of Public Service

[1] Paul Johnson interview, August 8, 2014.

Rent-a-Dad

[1] Goddard, 2014.

[2] Phil Gordon interview, August 21, 2014.

[3] Dan Campbell interview, November 18, 2013.

[4] Pat Graham interview, September 12, 2014.

[5] Goddard, 2014.

[6] Jack Pfister, "Glaring Lack of Civility in Hayworth's Letter," AR, December 7, 2003.

[7] John Fees interview, August 18, 2014.

[8] Paul Eckstein interview, July 5, 2013.

The Venice of Arizona

[1] Susan Herold, "Oasis: Arizona Canal Transformation Proposed," PG, February 26, 1988.

[2] Doug MacEachern, "Untapped," AR, July 12, 2009.

[3] Phyllis Gillespie, "Mayors Hail Plan for Canals," AR, April 27, 1990.

[4] Nan Ellin, emails to author, December 4 and 14, 2013.

The Core of Common Decency

[1] Suzanne Pfister, 2013.

[2] Avent, 2013.

[3] "@20: What I Wish I Knew Then," AR, April 20, 2007.

[4] Goddard, 2014.

[5] Suzanne Pfister, 2013.

[6] Saul Diskin interview, June 20, 2013.

[7] Carolyn Allen interview, June 1, 2013.

[8] Cathy Eden interview, January 28, 2014.

[9] Calderón, 2014.

[10] Johnny Cruz, "Former SRP Head and UA Alum Jack Pfister Dies," *UA News,* July 21, 2009.

INDEX

About the Author

Journalist Kathleen Ingley has specialized in covering Arizona's most complex and critical issues, including the state budget, water, urban development and energy. She was a reporter and editorial writer at the *Arizona Republic*, where her work included award-winning series on state trust land, the potential of solar energy, the threat of invasive plants, the increasing impact of the urban heat island and the challenge of growth ("An Acre an Hour: the Price of Sprawl"). She was business editor of the *Los Angeles Herald Examiner* and assistant editor of the *San Francisco Business Journal*. She has a bachelor's degree from the University of Michigan and a master's from the University of California at Berkeley. She served in the Peace Corps in Senegal.